Slow-Wave Microwave and mm-Wave Passive Circuits

Edited by

Philippe Ferrari
TIMA, Université Grenoble Alpes, CNRS, Grenoble INP, Grenoble, France

Anne-Laure Franc
LAPLACE, Université de Toulouse, CNRS, INPT, UPS, Toulouse, France

Marc Margalef-Rovira
STMicroelectronics, RFC HDC, Crolles, France

Gustavo P. Rehder
University of São Paulo, São Paulo, Brazil

Ariana Lacorte Caniato Serrano
University of São Paulo, São Paulo, Brazil

Registered Offices
John Wiley & Sons, Inc., 111 River Street, Hoboken, NJ 07030, USA
John Wiley & Sons Ltd, The Atrium, Southern Gate, Chichester, West Sussex, PO19 8SQ, UK

For details of our global editorial offices, customer services, and more information about Wiley products visit us at www.wiley.com.

Library of Congress Cataloging-in-Publication Data Applied for

Hardback ISBN: 9781119820161

Cover Design: Wiley
Cover Image: © Titima Ongkantong/Shutterstock

Set in 9.5/12.5pt STIXTwoText by Straive, Chennai, India
Printed and bound by CPI Group (UK) Ltd, Croydon, CR0 4YY

C9781119820161_061024

Contents

List of Contributors

Matthieu Bertrand
IMEP-LAHC, CNRS, Grenoble INP
Université Grenoble Alpes, Grenoble
France

Jordan Corsi
TIMA, Université Grenoble Alpes, CNRS
Grenoble INP, Grenoble, France

Philippe Ferrari
TIMA, Université Grenoble Alpes, CNRS
Grenoble INP, Grenoble, France

Anne-Laure Franc
LAPLACE, Université de Toulouse, CNRS
INPT, UPS, Toulouse, France

Leonardo Gomes
TIMA, Université Grenoble Alpes
Grenoble, France

and

Polytechnic School, University of São
Paulo, São Paulo, Brazil

and

STMicroelectronics, RFC HDC, Crolles
France

Hamza Issa
Faculty of Engineering, Beirut Arab
University, Beirut, Lebanon

Marc Margalef-Rovira
STMicroelectronics, RFC HDC, Crolles
France

Emmanuel Pistono
TIMA, Université Grenoble Alpes, CNRS
Grenoble INP, Grenoble, France

Gustavo P. Rehder
Polytechnic School, University of São
Paulo, São Paulo, Brazil

Abdelhalim Saadi
NXP, R&D HW RFP, Toulouse, France

Ariana Lacorte Caniato Serrano
Polytechnic School, University of São
Paulo, São Paulo, Brazil

Preface

Français

Écrire un livre sur les lignes à ondes lentes peut sembler une aventure périlleuse, car il y a mille et une manière de ralentir une onde électromagnétique. L'approche peut être basée sur l'utilisation de matériaux diélectrique ou magnétique, sur l'utilisation de composants tels des inductances ou des condensateurs, ou encore sur l'utilisation de méthodes de guidage particulières permettant de perturber les champs magnétique ou électrique. Toutes ces méthodes ont été étudiées au sein de la littérature scientifique.

Nombre d'entre elles ne sont pas abordées au sein de ce livre, en particulier deux méthodes très bien illustrées, à savoir l'utilisation de défauts de plan de masse et l'utilisation de composants localisés de type inductance, et surtout condensateur. L'utilisation de matériaux magnétique à forte perméabilité ou de matériaux diélectrique à forte constante diélectrique est également bien documentée au sein de la littérature, mais leur mise en oeuvre pose des problèmes, soit en termes de pertes, soit en termes de technologie, soit en termes de limitations électriques, en particulier par le fait qu'il est alors compliqué de réaliser des lignes de transmission possédant une impédance caractéristique au moins égale à 50 Ohms.

Après plus de quinze années de travaux de recherche au sein de nos équipes, nous, les cinq éditeurs de ce livre, avons décidé qu'il était temps de regrouper l'ensemble de nos travaux au sein d'un livre, afin de donner aux futurs lecteurs une vue globale de l'approche que nous avons développée. Cette approche consiste à construire des solutions de guidage des ondes en modifiant la topologie du champ électrique. Chacun aurait bien sûr le souhait de pouvoir également jouer avec le champ magnétique, mais la nature, qui souvent nous délivre des objets symétriques, n'a pas choisi la symétrie pour le comportement des champs électrique et magnétique. L'absence de charges magnétiques implique une distinction fondamentale des champs. Ainsi il s'avère très simple de configurer un champ électrique, en lui présentant les charges vers lesquelles il va naturellement se diriger. Le champ magnétique est plus sauvage, il tourne sur lui-même et ne peut être attiré, sauf par des matériaux de forte perméabilité, qui peuvent le dévier.

Sur la base de ce principe de configuration du champ électrique, nous présentons au sein de ce livre trois types de lignes à ondes lentes, (i) les lignes de type coplanaires, coplanar waveguide ou coplanar stripline, qui constituent le premier type de lignes à ondes lentes véritablement utilisées comme telles avec une très bonne efficacité en technologie intégrée de type CMOS, et mises en oeuvre pour la première fois par l'équipe de John Long, alors à TU Delft, puis (ii) les lignes microruban, et (iii) les guides d'ondes SIW. Sur la base de ces trois types de lignes, nous décrivons un grand nombre de circuits dont les performances et/ou les dimensions ont pu être améliorées grâce à l'utilisation du concept d'onde lentes.

In fine, nous espérons que les lecteurs prendront beaucoup de plaisir à parcourir ce livre que nous avons tous écrit avec grand plaisir.

English

Writing a book on slow-wave transmission lines may seem like a perilous adventure, as there are a thousand and one ways to slow down an electromagnetic wave. The approach is based on the use of dielectric or magnetic materials, the utilization of components such as inductors or capacitors, or even the application of specific guiding methods that disrupt the magnetic or electric fields. All of these methods have been studied within the scientific literature.

All these methods will not be addressed in this book, particularly only two methods are well illustrated: the use of defected ground planes and the use of lumped components such as inductors, especially capacitors. The use of high-permeability magnetic materials or high-dielectric constant dielectric materials is also well documented in the literature, but their implementation poses problems, either in terms of losses, technology, or electrical limitations. This is especially true because it becomes complicated to design transmission lines with a characteristic impedance of at least $50\,\Omega$ in such cases.

After more than 15 years of research within our teams, we, the five editors of this book, have decided that it is time to consolidate all of our work into a book to provide future readers with a comprehensive view of the approach we have developed. This approach involves constructing wave-guiding solutions by modifying the topology of the electric field. Everyone, of course, wishes to also play with the magnetic field, but nature, which often delivers symmetrical objects, has not chosen symmetry for the behavior of the electric and magnetic fields. The absence of magnetic charges implies a fundamental distinction in the fields. Thus, it turns out to be straightforward to configure an electric field by presenting charges toward which it will naturally move. The magnetic field is more untamed; it rotates on itself and cannot be attracted, except by materials with high permeability that can deflect it.

Based on this principle of electric field configuration, we present in this book three types of slow-wave structures: (i) coplanar lines (coplanar waveguide or coplanar stripline), which constitute the first type of slow-wave transmission lines genuinely used as such with very good efficiency in CMOS integrated technology, and first implemented by John Long's team at TU Delft, (ii) microstrip lines, and then (iii) substrate integrated waveguides (SIW). Building upon these three types of lines, we describe a large number of circuits whose performances and dimensions have been enhanced using the slow-wave concept.

In conclusion, we hope that readers will thoroughly enjoy reading this book, which we have all written with great pleasure.

Acronyms

3D	Three-Dimensional
AAO	Anodic Aluminum Oxide
AF-S-MS	Air-Filled Suspended S-MS
BEOL	Back-End-Of-Line
BiCMOS	Bipolar Complementary Metal-Oxide-Semiconductor
CAD	Computer-Aided Design
CCP	Crossed-Coupled Pair
CMOS	Complementary Metal-Oxide-Semiconductor
CPS	CoPlanar Stripline
CPW	CoPlanar Waveguide
DBR	Dual Behavior Resonator
DC	Direct Current
EM	Electromagnetic
FOM	Figure of Merit
FOMT	Figure of Merit with tuning range
FTR	Frequency Tuning Range
FWDC	Forward-Wave Directional Coupler
GSG	Ground Signal Ground
GSGSG	Ground Signal Ground Signal Ground
HR-SOI	High Resistivity Substrate On Insulator
IL	Insertion loss
LC	Liquid Crystals
LCP	Liquid Crystal Polymer
LNA	Low-Noise Amplifier
LRRM	Load Reflect Reflect Match
LTCC	Low Temperature Co-fired Ceramics
MEMS	MicroElectroMechanical System
mm-waves	Millimeter-waves
MOS	Metal-Oxide-Semiconductor
MOSFET	Metal Oxide Semiconductor Field-Effect Transistor
PA	Power Amplifier
PCB	Printed Circuit Board
PDK	Public Design Kit

QS	Quasi-Static
RF	RadioFrequency
RFIC	RadioFrequency Integrated Circuit
RL	Return Loss
SOI	Silicon on Insulator
S-MS	Slow-Wave Microstrip
SW	Slow-Wave
SWF	Slow-Wave Factor
TRL	Thru Reflect Line
VCO	Voltage-Controlled Oscillator
VNA	Vector Network Analyzer

1

Background Theory and Concepts

Philippe Ferrari[1], Marc Margalef-Rovira[2], and Gustavo P. Rehder[3]

[1] *TIMA, Université Grenoble Alpes, CNRS, Grenoble INP, Grenoble, France*
[2] *STMicroelectronics, RFC HDC, Crolles, France*
[3] *Polytechnic School, University of São Paulo, São Paulo, Brazil*

The objective of this introductory chapter is to draw up a brief history of slow-wave structures in Section 1.1, then to define the concept of slow-wave propagation from a theoretical point of view in Section 1.2, next to briefly present the three slow-wave transmission lines (in Section 1.3) presented in detail within the following chapters, namely slow-wave coplanar waveguides (S-CPWs, Chapter 2), slow-wave coplanar striplines (S-CPS, Chapter 2), slow-wave microstrip (S-MS, Chapter 3), slow-wave substrate integrated waveguides (SW-SIW, Chapter 4), and finally, in Section 1.4, to highlight some advantages of modern slow-wave transmission lines, which is a big part of the motivation for this book.

1.1 Historical Background

During the second half of the 20th century, technological advancements had greatly impacted modern society, with telecommunication networks being one of the most notable. With numerous scientific and technological breakthroughs, these networks have become more complex and efficient. This has led the consumer electronics market to become economically significant, with the development of new services and activities driven by the increasing demand for high-definition multimedia applications, secure data transmission, wearables, etc. To provide the necessary bandwidths and subsequent data rates for these applications, the next generation of wireless communications is oriented toward higher frequencies, especially the millimeter wave (mm-wave) bands. However, this presents new challenges, such as the need for wireless transceiver circuits that can operate at high frequencies with reasonable efficiencies, relying on low-cost technologies and compact solutions. Slow-wave structures can be a solution for the design of compact circuits.

The concept of slow-wave structures emerged in the early 1940s, for their capability to establish efficient interaction with electron beams. More precisely, it started in the context of radar applications where these interactions were used to amplify RF waves. The amplification was based on transferring kinetic energy from electrons to a propagating wave. Based

Slow-Wave Microwave and mm-Wave Passive Circuits, First Edition. Edited by Philippe Ferrari, Anne-Laure Franc, Marc Margalef-Rovira, Gustavo P. Rehder, and Ariana Lacorte Caniato Serrano.
© 2025 John Wiley & Sons Ltd. Published 2025 by John Wiley & Sons Ltd.

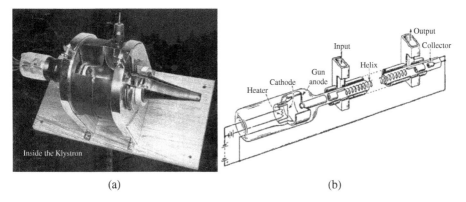

(a)   (b)

Figure 1.1 (a) First commercial klystron. Source: Henney 1940/with permission of WorldRadio-History; (b) Traveling-wave tube principle of operation. Source: Adapted from Kompfner (1947).

on this principle, the first slow-wave structure was called "klystron," a high-frequency vacuum tube invented in 1937 by W. Hansen and the Varian brothers (Varian & Varian, 1939), as illustrated in Fig. 1.1(a). In these amplifiers, the slow-wave propagation was achieved by cascading resonant cavities, which resulted in narrowband operation (Wu, 1999). In these structures, the slow-wave propagation was necessary because the interaction requires close velocities between the wave and the electrons, which move in vacuum at a lower velocity than the light. This interaction was further enhanced by R. Kompfner, who realized, in 1943, a broadband amplifier based on a nonresonant helix structure called "traveling-wave tube" (TWT; Kompfner, 1947), illustrated in Fig. 1.1(b). Improved versions of these devices are still in use for radar, satellite communications, television broadcasting, and particle accelerators. Oscillators have also been developed based on the same principles.

During the 1960s, the development of integrated microwave circuits provided a good opportunity to develop layered structures with potential slow-wave propagation. This concept was demonstrated for the first time in 1969 for a metal–insulator–semiconductor microstrip structure (Hasegawa et al., 1971) on silicon. It was followed by several topologies, including the Schottky contact transmission line (Jager, 1976). Thanks to an external bias, this last structure was used to create a variable slow-wave effect (Jaffe, 1972). As explained in (Wu, 1999), planar periodic structures gained attention in the early 1970s for the development of wide-band coupled microstrip lines (Podell, 1970). Since then, the research on planar periodic and layered structures has continued until today. In the meantime, slow-wave planar structures have also been used for miniaturization purpose. In microwave passive circuit design, specific functions such as filters, antennas, and couplers can be realized by the combination of physical phenomena such as interference, resonance, and couplings. These phenomena are very often dependent on wavelengths, so that specific properties can be obtained for given dimensions. It also means that, in general, these passive circuits occupy much larger areas than the active ones, which are made of increasingly smaller transistors. For example, a floating straight transmission line has intrinsic resonance frequencies that are directly related to the ratio of the propagation velocity and its physical length. Therefore, for a given frequency, a miniaturized structure can be obtained if the velocity is accordingly reduced. Obviously, the challenge lies in

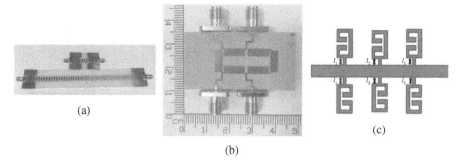

(a)

(b)

(c)

Figure 1.2 (a) Spoof surface plasmon-based slow-wave transmission lines. Source: Kianinejad et al. (2015)/with permission of IEEE; (b) Rat-race coupler. Source: Wei-Shin Chang et al. 2012/with permission of IEEE; (c) Six-resonator low-pass filter based on slow-wave resonators. Source: Shi et al. (2010)/with permission of IEEE.

the effort to make such slow-wave structures as efficient in terms of dissipation as the original ones. In printed-circuit-board technology, a high number of topologies have been developed in the recent years, some of them are illustrated in Fig. 1.2. Figure 1.2(a) shows spoof surface plasmon-based transmission lines, for compact designs, reduced attenuation and limited cross-talk (Kianinejad et al., 2015). Compact couplers, such as the "rat-race," are illustrated in Fig. 1.2(b). The compactness was achieved by using a high slow-wave factor (SWF) microstrip structure (Chang & Chang, 2012). In Fig. 1.2(c), a filter based on six slow-wave resonators is also shown (Shi et al., 2010). One could also mention the use of defected ground (Kim & Lee, 2006) and electromagnetic band-gap (Zhurbenko et al., 2006), which often exhibit slow-wave propagation.

Concerning integrated technologies, the design of compact and low-loss passive circuits is a real challenge. It is especially true for the newly addressed mm-wave bands, in which parasitic couplings are more and more limiting and where the high conduction losses result in poor quality factors. In this context, slow-wave structures do not only provide miniaturized circuits but may also lead to higher quality factors (Chee et al., 2006; Cheung & Long, 2006; Franc et al., 2013). This is the case of the S-CPW, whose geometry prevents conduction losses in the semiconductor by shielding the electric field (see Fig. 1.3(a)). Meander lines were also used to realize compact couplers in silicon-based integrated passive device (IPD) technologies (Tseng & Chen, 2016), as shown in Fig. 1.3(b). A band-pass filter using a slow-wave microstrip topology was presented in (Evans et al., 2012), it is illustrated in Fig. 1.3(c).

1.2 The Slow-Wave Concept

In this section, a general theoretical approach explaining the concept of slow-wave propagation is presented. It is based on the magnetic and electric energies that are stored in a waveguide (Bertrand et al., 2020).

A general uniform waveguide topology is illustrated in Fig. 1.4.

The cross section could contain either different metallic conductors, magnetic, or dielectric materials. A wave propagating inside such a waveguide is characterized by its phase

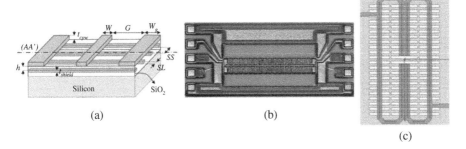

Figure 1.3 (a) Slow-wave coplanar waveguide topology. Source: Franc et al. (2013)/with permission of IEEE; (b) Slow-wave coupler in silicon-based IPD technology. Source: Tseng et al. 2016/with permission of IEEE; (c) Miniaturized slow-wave microstrip filter in 65-nm CMOS technology. Source: Evans et al. (2012)/with permission of The Institution of Engineering and Technology.

constant β and angular frequency ω. By definition, its phase velocity v_p is the velocity at which the phase of the wave travels in space, and is defined as (1.1).

$$v_p = \frac{\omega}{\beta} \tag{1.1}$$

It can be seen as the velocity at which an observer should travel along the waveguide in order to keep in state with this wave. A second velocity is called group velocity, v_g, and it is the velocity at which the overall shape of the waves' amplitudes – or modulation – travels through space. This velocity is also often interpreted as the velocity at which the energy or information propagates; it is given by (1.2).

$$v_g = \frac{\partial \omega}{\partial \beta} \tag{1.2}$$

By definition, for a non-dispersive propagating mode such as the lossless TEM (Transverse Electric Magnetic) mode, the phase velocity does not depend on frequency, which implies that the phase constant β is a linear function of ω. In that case, it follows that $v_p = v_g$. However, if this velocity does vary, the group velocity differs from the phase velocity. The phase velocity (and therefore the group velocity) in a specific material is fixed by its permittivity ε and permeability μ, in particular $v_p = v_g = c = 1/\sqrt{\varepsilon \cdot \mu}$. For a guided wave, however, dispersion occurs so that phase and group velocities can greatly differ from the free-space value. As explained in (Wu, 1999), in a closed uniform waveguide, waves propagate at velocities greater than the light velocity; they are commonly called "fast-waves." This is equivalent to say that the guided wavelength λ is greater than the free-space wavelength. On the contrary, some structures can exhibit lower velocities compared to free-space propagation. These waves are therefore called "slow-waves" and are characterized by smaller guided wavelengths. The velocity reduction can be obtained by specific spatial variations of material properties in the transverse section, as shown in Fig. 1.4. Another method relies on the introduction of periodicity in the propagation direction, either in material properties or in the boundary conditions. In order to quantify the velocity reduction, a common definition of a so-called *swf* is adopted (Wu, 1999). This factor is defined as the ratio between the free-space velocity in vacuum c_0 and actual phase velocity in the considered waveguide v_p (see (1.3)). This factor is also called the effective refractive index, as defined in optics.

Figure 1.4 General form of a uniform waveguide.

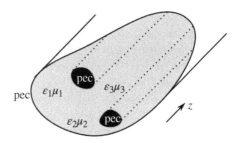

$$swf = \frac{c_0}{v_p} = \frac{\lambda_0}{\lambda} = \frac{\beta}{\beta_0} = \sqrt{\varepsilon_{reff}} \tag{1.3}$$

It can also be expressed as the ratio of free-space and guided wavelengths λ_0 and λ, or phase constants β_0 and β, respectively. Usually, an effective relative permittivity ε_{reff} is introduced, integrating both magnetic and electrical effects into one single parameter (as in Wu, 1999). In this definition, ε_{reff} contains the contribution of material properties (ε_r, μ_r), as well as any additional slow-wave mechanism.

From the definition given previously, a slow-wave structure is characterized by a *swf* greater than one. However, based on this definition, most existing waveguides (TEM or not) are slow-wave structures because they contain dielectric materials with $\varepsilon_r > 1$. For convenience, a second definition of SWF may be used to remove this ambiguity. It is defined as the ratio between the phase velocity in a given wave-guiding structure v_p^{ref} taking the material properties into account, and v_p, the one achieved through additional geometry and material modifications (see (1.4)).

$$SWF = \frac{v_p^{ref}}{v_p} = \frac{\lambda^{ref}}{\lambda} = \frac{\beta}{\beta^{ref}} \tag{1.4}$$

Besides, if the reference waveguide supports a TEM mode so that its phase velocity can be expressed as $v_p^{ref} = c_0/\sqrt{\varepsilon_r}$, then the following relation between the two definitions can be derived:

$$SWF = \frac{v_p^{ref}}{v_p} = \frac{swf}{\sqrt{\varepsilon_r}} \tag{1.5}$$

As a result, the effective relative permittivity becomes:

$$\varepsilon_{reff} = \varepsilon_r \cdot SWF^2 \tag{1.6}$$

The scientific literature related to the design of slow-wave waveguides and passive components regularly claims that the slow-wave propagation is conditioned by a spatial separation of electric and magnetic energies (Wu, 2005). This separation requirement is generally mentioned along with the necessity to increase stored energy in the structure (Jin et al., 2016; Niembro-Martin et al., 2014). Naturally, the impact of the field distribution modification within slow-wave structures is usually described in terms of increased capacitance or inductance using distributed L-C models (Bastida & Donzelli, 1979).

A general explanation of the relationship between spatial separation and reduced propagation velocity without recurring to circuit approximate modeling is provided further. One can distinguish between two approaches in generating slow-wave propagation. The first one relies on using specific material arrangement in the transverse section of a longitudinally uniform waveguide. These materials may include dielectrics with different dielectric constants or more complex materials, such as semi-conductors, for instance. The second approach relies on the introduction of longitudinally periodic variations of boundary conditions or material properties. It is worth noting that for lossless periodic waveguides, allowing propagation in specific frequency pass-band, stored electric and magnetic energies obey the two theorems as follows (Collin, 1990; Watkins, 1958):

Theorem 1.1 *The time-average stored electrical energy per period is equal to the time-average stored magnetic energy per period in the pass-band.*

Theorem 1.2 *The time-average power flow in the pass-bands is equal to the group velocity times the time-average stored electrical and magnetic energy per period divided by the period.*

These two theorems can be expressed by equations (1.7) and (1.8), respectively, using the field complex vectors $\boldsymbol{E}$ and $\boldsymbol{H}$.

$$\iiint_V \frac{1}{4}\varepsilon|\boldsymbol{E}|^2 dV - \iiint_V \frac{1}{4}\mu|\boldsymbol{H}|^2 dV = W_e - W_m = 0 \tag{1.7}$$

$$\iint_{S_t(z)} \frac{1}{2}\Re(\boldsymbol{E} \times \boldsymbol{H}^*) \cdot d\boldsymbol{S} = \left[\iiint_V \left(\frac{1}{4}\varepsilon|\boldsymbol{E}|^2 + \frac{1}{4}\mu|\boldsymbol{H}|^2\right) dV\right] \frac{v_g}{s} \tag{1.8}$$

where V is the volume of an s-periodic waveguide period along z coordinate (propagation direction), $S_t(z)$ its transverse section, and v_g the group velocity. We also denote the stored electric and magnetic energies in a period as W_e and W_m, respectively. It is important to note that in the case of uniform waveguides, these two theorems also apply, as the period s can be fixed to any real value. From these two theorems, it naturally follows that for two waveguides, namely 1 and 2, having identical periods and carrying the same amount of power, the energy stored and the group velocities are closely related (1.9).

$$W_{e1}v_{g1} = W_{m1}v_{g1} = W_{e2}v_{g2} = W_{m2}v_{g2} \tag{1.9}$$

In other words, the ratio of group velocities for propagation in the aforementioned waveguides is directly given by the ratios of stored energies, see (1.10).

$$\frac{v_{g1}}{v_{g2}} = \frac{W_{e2}}{W_{e1}} = \frac{W_{m2}}{W_{m1}} \tag{1.10}$$

Now, in the case of a low dispersion propagation scheme, this ratio can be expressed as the ratio of phase velocities, as shown in (1.11).

$$\frac{v_{p1}}{v_{p2}} \approx \frac{W_{e2}}{W_{e1}} = \frac{W_{m2}}{W_{m1}} \tag{1.11}$$

Most slow-wave waveguides in the literature consist of modified versions of well-known wave-guiding topologies, such as closed circular or rectangular waveguides or planar transmission lines (coplanar, microstrip). Therefore, it is usually convenient to use such an existing waveguide as point of reference when dealing with the performance of a specific slow wave implementation.

As a conclusion, the phase velocity reduction in slow-wave structures can be related to energy considerations, if both low dispersion and low losses are assumed. In this case, the reduction in velocity as compared to a reference waveguide carrying the same amount of power is approximated by the equation (1.12).

$$SWF \approx \frac{W_e^{ref}}{W_e} = \frac{W_m^{ref}}{W_m}$$

(1.12)

where W_e and W_m are the time-averaged energies stored in a period of the slow-wave waveguide (or arbitrary section if uniform), while W_e^{ref} and W_m^{ref} correspond to the chosen reference waveguide.

1.3 Modern Slow-Wave Transmission Lines Brief Description

As highlighted in Section 1.1, the concept of slow-wave transmission lines is not new in the 21st century, but the development of slow-wave transmission lines making it possible to produce RF and mm-wave circuits with high efficiency is relatively new (early 2000s for the first demonstrations). It is in this sense that we use the word "modern" to qualify the slow-wave transmission lines presented in this book.

1.3.1 Slow-Wave Coplanar Waveguide

As shown in Fig. 1.5, the S-CPW is simply a CPW under which a floating shield has been inserted. The electrically floating shield consists of metallic ribbons placed in a perpendicular configuration, as compared to the direction of propagation. The electrical and magnetic field lines for the central conductor are drawn in Fig. 1.5. As it will be explained in detail in Chapter 2, the magnetic field, as compared to a classical CPW, is almost unperturbed by the floating shield while the electric field is captured by it.

From a circuit point of view, this leads to an increase in the CPW linear capacitance C_{lin} and, hence, a decrease in the phase velocity v_p, which is given by equation (1.13):

$$v_p = \frac{1}{\sqrt{L_{lin} \cdot C_{lin}}}$$

(1.13)

with L_{lin} the linear inductance of the CPW. The decrease in phase velocity leads to a reduction in the physical length of the transmission lines, for a given electrical length, as well as

Figure 1.5 S-CPW 3D view.

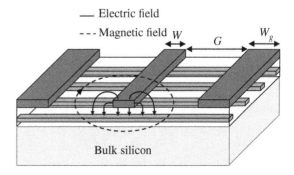

an improvement in electrical performance with regard to integrated technologies on lossy substrates, e.g. silicon, as developed in Chapter 2, since it prevents the electric field from reaching the substrate.

1.3.2 Slow-Wave Microstrip (S-MS)

The concept of S-MS was first proposed by Coulombe et al. (2007), with the first equivalent electrical model being given by Serrano et al. (2014). The topology of the S-MS is illustrated in Fig. 1.6.

As for the CPW, the S-MS is simply a microstrip line under which a so-called forest of vertical metallic posts has been inserted. From an electrical point of view, the effect of the forest of metallic posts is the same as that of the floating shield of the S-CPW. The electrical and magnetic field lines are drawn in Fig. 1.6.

As will be explained in detail in Chapter 3, the magnetic field passes through the forest of metallic posts, while the electric field is captured by them. From a circuit point of view, this leads to an increase in the microstrip line linear capacitance C_{lin} and, hence, a decrease in the phase velocity v_p.

1.3.3 Slow-Wave Substrate Integrated Waveguide (SW-SIW)

The concept of SW-SIW was first proposed by Niembro-Martin et al. (2014). The topology of the SW-SIW is given in Fig. 1.7.

The same concept used for the S-MS, namely a forest of metallic posts, is used to build a SW-SIW, where the posts capture the electric field, and the magnetic field is hardly modified. The electric field is captured by the posts; thus, a current is flowing through the posts to the bottom metal. This current creates a magnetic field around each post. This magnetic field is thus present in the whole forest of posts. From a circuit point of view, this leads to a decrease in the SIW cut-off frequency and phase velocity v_p. However, as explained in detail in Chapter 4, the magnetic field within the volume of the forest of posts is created by the current flowing in these, while the magnetic field in the upper volume, above the vias, is of course created by the variable electric field. This behavior is given by the Maxwell-Ampere equation, see (1.14).

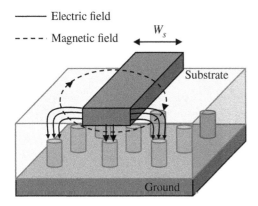

——— Electric field

- - - - Magnetic field

W_s

Substrate

Ground

Figure 1.6 3D illustration of an S-MS and its magnetic and electric fields.

—— Electric field

--- Magnetic field

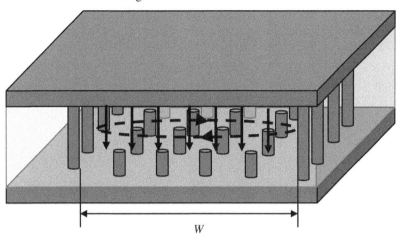

Figure 1.7 SW-SIW 3D view.

$$\vec{\nabla} \times \vec{H} = \vec{J} + \varepsilon \frac{\partial \vec{E}}{\partial t} \tag{1.14}$$

where $\vec{H}$ the magnetic field, $\vec{E}$ the electric field, $\vec{J}$ the current density, and ε the dielectric constant. Considering equation (1.14), the magnetic field in the upper volume (i.e. without vias) is given by the second part of the term on the right, as in equation (1.15).

$$\vec{\nabla} \times \vec{H} = \varepsilon \frac{\partial \vec{E}}{\partial t} \tag{1.15}$$

whereas the magnetic field in the bottom volume (i.e. with posts) is given by the first part of the term on the right, as in equation (1.16).

$$\vec{\nabla} \times \vec{H} = \vec{J} \tag{1.16}$$

1.4 Motivations for the Development of Modern Slow-Wave Transmission Lines

In this section, we give the major arguments that have motivated our research on slow-wave transmission lines since the 2000s. The major motivations for the writing of this book are the following:

- Improvement of transmission line performance with integrated technologies;
- Reduction of the transmission lines and SIWs length;
- Add new degrees of freedom in the development of coupled lines and 3D transmission lines.

1.4.1 Improvement of Transmission Lines Performance in Integrated Technologies

In the first proposals of integrated transmission lines, the innovative concept consisted of adding electrically floating metallic ribbons, forming a floating shield under a coplanar waveguide (CPW). From a field point of view, these ribbons separate the electric and magnetic fields under them, where only the magnetic field could flow. From a circuit point of view, the floating shield can be considered as a loading capacitance that is distributed along the CPW, thus leading to a slow-wave. Next, keeping in mind that the quality factor Q of a transmission line can be expressed as half of the ratio between the propagation constant β and the attenuation constant α, as in equation (1.17).

$$Q = \frac{1}{2}\frac{\beta}{\alpha} \tag{1.17}$$

Increasing β without modifying α leads to an improvement in the quality factor (Q). This is the case in S-CPWs in integrated technologies, for which the attenuation constant is kept almost constant while increasing the propagation constant due to the decreases in phase velocity (equation (1.18)).

$$\beta = \frac{\omega}{v_p} \tag{1.18}$$

Next, except for interconnection purposes, the most important point for a designer is not the physical length of a transmission line but rather its electrical length, and its characteristic impedance. If the phase velocity v_p decreases, the guided wavelength λ_g decreases for a given operating frequency f, equation (1.19). Then, thanks to the slow-wave effect, the physical length l is shorter for the same electrical length θ, equation (1.20).

$$v_p = \lambda_g \cdot f \tag{1.19}$$

$$\theta = \frac{\omega}{v_p} \cdot l \tag{1.20}$$

In order to properly consider this effect, we can rewrite the equation (1.17) as expressed in equation (1.21).

$$Q = \frac{1}{2}\frac{\beta \cdot l}{\alpha \cdot l} = \frac{8.68}{2}\frac{\theta}{IL_{dB}} = 4.34\frac{\theta}{IL_{dB}} \tag{1.21}$$

with IL_{dB} being the insertion loss in dB. Hence, the insertion loss of a piece of transmission line of electrical length θ can be simply expressed using the quality factor (equation (1.22)).

$$IL_{dB} = 4.34\frac{\theta}{Q} \tag{1.22}$$

Hence, when Q increases, IL_{dB} decreases for a given electrical length θ. Note that, contrary to integrated technologies, slow-wave transmission lines do not lead to quality factor improvement in PCB technologies. The reason is explained further. In integrated technologies, only microstrip lines and CPWs (and their "cousin" coplanar striplines, CPS) can be considered for most practical cases. Their transversal views are given in Fig. 1.8. Normally, the microstrip is implemented on the Back-End-Of-Line (BEOL) using the top metal for the signal strip and the bottom metal (sometimes a stack of two metals) for the ground plane. In

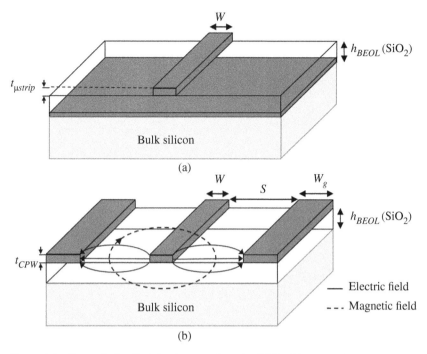

Figure 1.8 Transmission lines implemented on the BEOL of integrated technologies: (a) Microstrip; (b) CPW.

this case, the substrate "seen" by the electrical field is the SiO_2 between the top and bottom metals. In CPW, the top metal, which is normally the thicker metal, is used.

Due to the reduced thickness of the BEOL of the current technologies, i.e. around 5 μm for CMOS (complementary metal-oxide-semiconductor) technologies and around 10 μm for advanced BiCMOS (Bipolar-CMOS) technologies, the quality factor of microstrip lines is limited to approximately 10–20 near 60 GHz, due to the limited width of the signal strip W for a characteristic impedance near 50 Ω. CPWs make it theoretically possible to overcome this problem, by offering more flexibility in the adjustment of the dimensions (gap S and strip W), independently of the substrate thickness. Nevertheless, the conductivity of the bulk silicon substrate leads to excessive losses when the gap is too large (i.e. larger than the BEOL thickness), thus drastically limiting the use of CPWs. Even with the reduced height of the BEOL, the microstrip lines have equal or even better quality factors than CPW, with better compactness and easier use, in particular when T-junction must be designed. To illustrate this, CPWs of different dimensions were simulated with ANSYS HFSS (High Frequency Simulation Software). The extraction of the attenuation constant α, relative effective dielectric constant ε_{reff}, and quality factor Q are given in Figs. 1.9–1.11, and 1.12, respectively. These simulations were carried out for BEOLs of thickness 5 and 10 μm, respectively, and top metal thicknesses of 1 and 2 μm, respectively. Standard bulk silicon resistivity of 10 Ω · cm was considered. It can be considered that a BEOL thickness of 5 μm associated with a top metal thickness of 1 μm is a CMOS-like technology, whereas a BEOL thickness of 10 μm associated with a top metal thickness of 2 μm is closer to a BiCMOS-like technology.

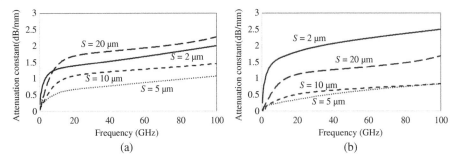

Figure 1.9 Attenuation constant of 50-Ω characteristic impedance CPWs (ANSYS EM Suite). (a) BEOL thickness 5 μm and strip thickness 1 μm (CMOS-like). (b) BEOL thickness 10 μm and strip thickness 2 μm (BiCMOS-like).

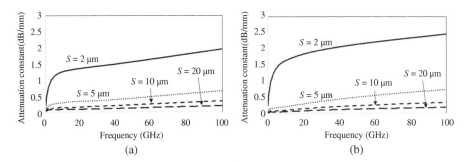

Figure 1.10 Attenuation constant of 50-Ω characteristic impedance CPWs (HFSS simulations) with perfect bulk silicon (infinite resistivity). (a) BEOL thickness 5 μm and strip thickness 1 μm (CMOS-like). (b) BEOL thickness 10 μm and strip thickness 2 μm (BiCMOS-like).

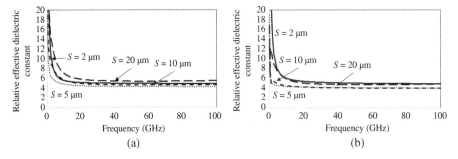

Figure 1.11 Effective dielectric constant of 50-Ω characteristic impedance CPWs (HFSS simulations). (a) BEOL thickness 5 μm and strip thickness 1 μm (CMOS-like). (b) BEOL thickness 10 μm and strip thickness 2 μm (BiCMOS-like).

This terminology is used as follows. Moreover, 50-Ω characteristic impedance CPWs were considered. The gap S of the CPW is given in the figures. Note that the signal strip widths W is not the same for CMOS-like and BiCMOS-like technologies, due to the different metal thicknesses. For a given gap S, the linear capacitance C of the CPW increases when the metal thickness increases, in particular for small gaps, as illustrated in Fig. 1.13. To

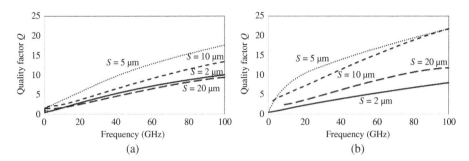

Figure 1.12 Quality factor Q of 50-Ω characteristic impedance CPWs (HFSS simulations). (a) BEOL thickness 5 μm and strip thickness 1 μm (CMOS-like). (b) BEOL thickness 10 μm and strip thickness 2 μm (BiCMOS-like).

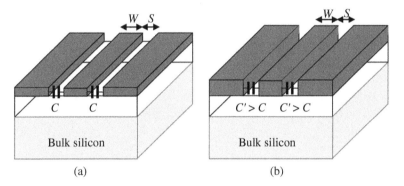

Figure 1.13 CPW linear capacitance C. The thickness of the metals is 1 μm (a) and 2 μm (b), respectively. When the gap S becomes very small, of the order of a few μm, the thickness of the metals plays an important role, the "parallel plate" capacitance becoming predominant. Thus, the linear capacitance C increases versus metal thickness.

compensate for the increase in C, the linear inductance L must be increased in order to maintain a characteristic impedance Z_c equal to 50 Ω, since $Z_c = \sqrt{L/C}$. The increase in L can only be achieved by reducing the width of the signal strip, W. The consequence is an increase in the linear resistance R of the CPW, and therefore an increase in the attenuation constant. Thus, for very small gaps S, in particular 2 μm, the attenuation constant is higher for the BiCMOS-like technology, as compared to the CMOS-like technology, even though the metals are thicker. This can be observed in Fig. 1.9 and Fig. 1.10, which represent the attenuation constant, considering a conventional bulk silicon in Fig. 1.9 (resistivity 10 $\Omega \cdot$ cm) and an ideal bulk silicon in Fig. 1.10 (infinite conductivity), respectively. For $S > 5$ μm, the interest of a thicker BEOL as well as a thicker conductor is clearly seen, resulting in a lower attenuation constant. Fig. 1.14 shows the W/S ratio for the two types of BEOL considered, illustrating the fact that for small S, W becomes smaller for thicker metals. Finally, Fig. 1.10 shows that the attenuation constant will continue to decrease as S increases with perfect bulk silicon. Hence, high-resistivity-type substrate makes it possible to improve the quality factor of the CPWs, but this is complicated and expensive to implement in the context of conventional circuits with a large number of transistors.

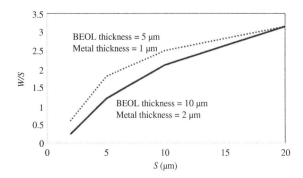

Figure 1.14 *W/S* ratio for the simulation results presented in Figs. 1.9–1.12.

The effective dielectric constant ε_{reff} is given in Fig. 1.11. Apart from low frequency below 20 GHz, ε_{reff} is relatively constant and not very dependent on S. Finally, the quality factor Q is given in Fig. 1.12. Q reaches its maximum for small gaps, around 5 μm, for both technologies, CMOS- and BiCMOS-like, with values of 13 and 17 at 60 GHz for CMOS-like and BiCMOS-like technologies, respectively. Hence, CPW and microstrip lines exhibit almost the same quality factor, from 10 to 20 around 60 GHz. It will be shown in Chapter 2 that much higher Q can be reached with S-CPW.

Finally, note also that it is not possible to use striplines (see Fig. 1.15(a)), because (i) the BEOL offered by the technologies is very thin, with only the top metal layers being thick, and (ii) metal thicknesses decrease from the top to the bottom of the BEOL. Hence, the design of a stripline would suffer from two intolerable issues leading to very poor performance: (i) a dissymmetry, with top ground to strip height different from strip to bottom ground, and (ii) thick top ground and thin to very thin strip and bottom ground, as shown in Fig. 1.15(b).

Things are very different in PCB technology because the thickness of the substrates is of the order of a few hundred microns. Fig. 1.16 gives the frequency of occurrence of the first higher-order mode of a microstrip line as a function of the substrate thickness. A substrate thickness of 700 μm can be used up to 40 GHz and 200 μm up to more than 120 GHz. In addition, there is no equivalence to the bulk silicon, which could lead to extra losses. As a consequence, the quality factor of microstrip lines reaches more than 150 and 270 at mm-waves above 30 GHz, as shown in Fig. 1.17(a), for a loss tangent equal to 0.005 and 0.002, respectively. The losses are mainly dielectric, as shown in Fig. 1.17(b). Results would be comparable for CPW in PCB technology.

Adding metallic posts or floating ribbons to microstrip lines and CPWs, in order to realize S-MS or S-CPW in PCB technology, would result in lower quality factor, since losses are mainly dielectric, unlike CMOS/BiCMOS technologies, where losses are mainly metallic and related to the semiconductor substrate. Due to the BEOL thickness limitation

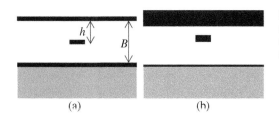

Figure 1.15 Stripline topology. (a) Ideal stripline, with symmetry. (b) Unsymmetrical stripline that could be realized in a standard BEOL.

(a)   (h)

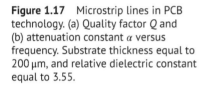

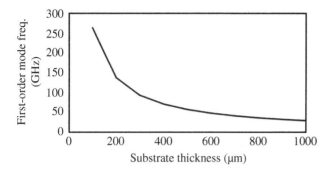

Figure 1.16 Frequency of occurrence of the first higher-order mode of a microstrip line as a function of the substrate thickness. A relative dielectric constant equal to 3.55 was considered.

Figure 1.17 Microstrip lines in PCB technology. (a) Quality factor Q and (b) attenuation constant α versus frequency. Substrate thickness equal to 200 µm, and relative dielectric constant equal to 3.55.

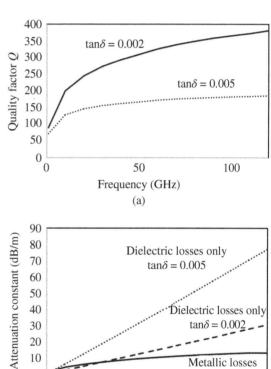

in CMOS/BiCMOS technologies, it is interesting to use the flexibility offered by CPW to enlarge the strips width, which is efficient, thanks to bulk silicon shielding carried out by the slow-wave approach, but this approach loses all its interest if the thickness of the substrate is large, as is the case with PCB technology. To conclude, S-CPS or S-CPW are interesting in CMOS/BiCMOS technologies in order to improve performance, but not in PCB technology, where microstrip lines or CPW lead to a higher quality factor.

1.4.2   Reduction of the Transmission Lines and SIWs Length

The slow-wave effect makes it possible to reduce the length of the transmission lines and the SIWs, whether in PCB or integrated technologies. For transmission lines, i.e. S-CPWs and S-MS lines, this comes from the reduction in phase velocity, v_p, given in equation (1.13). Since the electrical length θ is inversely proportional to the phase velocity (equation (1.20)), decreasing v_p increases θ for a given physical length l. Or conversely, for a given electrical length, the physical length is reduced. We can also analyze things using the effective relative dielectric constant, which increases with the slow-wave effect, as shown in equations (1.23) and (1.24).

$$v_p = \frac{C_0}{\sqrt{\varepsilon_{reff}}} \tag{1.23}$$

$$\varepsilon_{reff} = C_0(L_{lin} \cdot C_{lin})^2 \tag{1.24}$$

with L_{lin} and C_{lin} the linear inductance and linear capacitance of the transmission line.

For SIWs, in the same way, the effective dielectric constant increases, which reduces its cut-off frequency and therefore increases the electrical length at a given frequency and for a given physical length. A particularity of SIWs is linked to the fact that the slow-wave effect, as described in this book, also makes it **possible to reduce the lateral dimensions of the waveguides**, as will be explained in Chapter 4.

1.4.3   Addition of New Degrees of Freedom in the Development of Coupled-Lines and 3D Transmission Lines

The necessary vias/ribbons to achieve the slow-wave effect are an additional element to control the response of the structure. Hence, the use of slow-wave architectures does not only present advantages in terms of miniaturization and Q-factor improvement as introduced before but also increases design flexibility. For instance, the characteristic impedance in a CPW is controlled by the gap between the signal and ground strips, and the width of the former. On the other hand, in a S-CPW, the impedance can additionally be controlled by the density of ribbons below the structure and the height at which they are placed. The floating ribbons can also be connected to the ground strips or not. This feature is extremely interesting when dealing with structures holding multiple modes of propagation, as they can be controlled independently. This was exploited for the design of the coupled slow-wave coplanar waveguides (CS-CPW), firstly presented in Lugo-Alvarez et al. (2014), whose structure is presented in Fig. 1.18. More precisely, in this architecture, the use of the floating ribbons, with the additional possibility of performing cuts on them, allows to independently control the even- and odd-modes.

Moreover, the control over the spatial distribution of the electromagnetic field achieved thanks to the slow-wave effect, permits the design of extremely compact architectures in a three-dimensional approach. With this view, the Folded S-CPW (FS-CPW) was proposed in Margalef-Rovira et al. (2019). This structure, shown in Fig. 1.19, aims at stacking two (or more) S-CPWs to increase the compactness of the transmission line and can lead to very compact resonators and filters.

Hence, the slow-wave structures do not only allow to increase the compactness and Q-factor of transmission lines, but also to present additional degrees of freedom and permit 3D designs, especially interesting in integrated technologies.

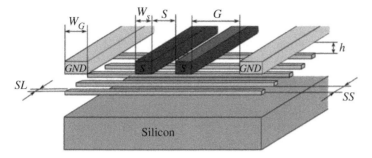

Figure 1.18 Conceptual view of the CS-CPW architecture.

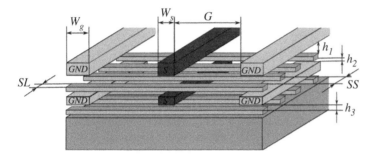

Figure 1.19 Conceptual view of the FS-CPW architecture.

References

Bastida, E. M., & Donzelli, G. P. (1979). Periodic slow-wave low-loss structures for monolithic GaAs microwave integrated circuits. *Electronics Letters*, 15(19), 581. https://doi.org/10.1049/el:19790417

Bertrand, M., Corsi, J., Pistono, E., Kaddour, D., Puyal, V., & Ferrari, P. (2020). On the effect of field spatial separation on slow wave propagation. *IEEE Transactions on Microwave Theory and Techniques*, 68(12), 4978–4983. https://doi.org/10.1109/TMTT.2020.3023445

Chang, W. S., & Chang, C. Y. (2012). A high slow-wave factor microstrip structure with simple design formulas and its application to microwave circuit design. *IEEE Transactions on Microwave Theory and Techniques*, 60(11), 3376–3383. https://doi.org/10.1109/TMTT.2012.2216282

Chee, I., Lai, H., Tanimoto, H., & Fujishima, M. (2006). Characterization of high Q transmission line structure for advanced CMOS processes. *IEICE Transactions on Electronics*, E89-C(12), 1872–1879. https://doi.org/10.1093/ietele/e89-c.12.1872

Cheung, T. S. D., & Long, J. R. (2006). Shielded passive devices for silicon-based monolithic microwave and millimeter-wave integrated circuits. *IEEE Journal of Solid-State Circuits*, 41(5), 1183–1200. https://doi.org/10.1109/JSSC.2006.872737

Collin, R. E. (1990). *Field Theory of Guided Wave* (Vol. 5). John Wiley & Sons.

Coulombe, M., Nguyen, H. V., & Caloz, C. (2007). Substrate integrated artificial dielectric (SIAD) structure for miniaturized microstrip circuits. *IEEE Antennas and Wireless Propagation Letters*, 6, 575–579. https://doi.org/10.1109/LAWP.2007.910959

Evans, R. J., Skafidas, E., & Yang, B. (2012). Slow-wave slot microstrip transmission line and bandpass filter for compact millimetre-wave integrated circuits on bulk complementary metal oxide semiconductor. *IET Microwaves, Antennas & Propagation*, 6(14), 1548–1555. https://doi.org/10.1049/iet-map.2012.0336

Franc, A.-L., Pistono, E., Meunier, G., Gloria, D., & Ferrari, P. (2013). A Lossy Circuit Model based on physical interpretation for integrated shielded slow-wave CMOS coplanar waveguide structures. *IEEE Transactions on Microwave Theory and Techniques*, 61(2), 754–763. https://doi.org/10.1109/TMTT.2012.2231430

Hasegawa, H., Furukawa, M., & Yanai, H. (1971). Properties of microstrip line on Si-SiO/ sub 2/System. *IEEE Transactions on Microwave Theory and Techniques*, 19(11), 869–881. https://doi.org/10.1109/TMTT.1971.1127658

Henney, K. (1940). Inside the klystron. *Electronics Magazine*. cover.

Jaffe, J. M. (1972). A high-frequency variable delay line. *IEEE Transactions on Electron Devices*, 19(12), 1292–1294. https://doi.org/10.1109/T-ED.1972.17593

Jager, D. (1976). Slow-wave propagation along variable Schottky-Contact microstrip line. *IEEE Transactions on Microwave Theory and Techniques*, 24(9), 566–573. https://doi.org/10.1109/TMTT.1976.1128910

Jin, H., Wang, K., Guo, J., Ding, S., & Wu, K. (2016). Slow-wave effect of substrate integrated waveguide patterned with microstrip polyline. *IEEE Transactions on Microwave Theory and Techniques*, 64(6), 1717–1726. https://doi.org/10.1109/TMTT.2016.2559479

Kianinejad, A., Chen, Z. N., & Qiu, C.-W. (2015). Design and modeling of spoof surface plasmon modes-based microwave slow-wave transmission line. *IEEE Transactions on Microwave Theory and Techniques*, 63(6), 1817–1825. https://doi.org/10.1109/TMTT.2015.2422694

Kim, H.-M., & Lee, B. (2006). Bandgap and slow/fast-wave characteristics of defected ground structures (DGSs) including left-handed features. *IEEE Transactions on Microwave Theory and Techniques*, 54(7), 3113–3120. https://doi.org/10.1109/TMTT.2006.877060

Kompfner, R. (1947). The traveling-wave tube as amplifier at microwaves. *Proceedings of the IRE*, 35(2), 124–127. https://doi.org/10.1109/JRPROC.1947.231238

Lugo-Alvarez, J., Bautista, A., Podevin, F., & Ferrari, P. (2014). High-directivity compact slow-wave CoPlanar waveguide couplers for millimeter-wave applications. *2014 44th European Microwave Conference*, 1072–1075. https://doi.org/10.1109/EuMC.2014.6986624

Margalef-Rovira, M., Occello, O., Saadi, A. A., Barragan, M. J., Gaquiere, C., Pistono, E., Bourdel, S., & Ferrari, P.2019). A 30-GHz compact resonator structure based on folded slow-wave CoPlanar waveguides on a 55-nm BiCMOS technology. *7th Advanced Electromagnetics Symposium (AES)*.

Niembro-Martin, A., Nasserddine, V., Pistono, E., Issa, H., Franc, A.-L., Vuong, T.-P., & Ferrari, P. (2014). Slow-wave substrate integrated waveguide. *IEEE Transactions on Microwave Theory and Techniques*, 62(8), 1625–1633. https://doi.org/10.1109/TMTT.2014.2328974

Podell, A. (1970). A high directivity microstrip coupler technique. *G-MTT 1970 International Microwave Symposium*, 33–36. https://doi.org/10.1109/GMTT.1970.1122761

Serrano, A. L. C., Franc, A.-L., Assis, D. P., Podevin, F., Rehder, G. P., Corrao, N., & Ferrari, P. (2014). Modeling and characterization of slow-wave microstrip lines on metallic-nanowire-filled-membrane substrate. *IEEE Transactions on Microwave Theory and Techniques*, 62(12). https://doi.org/10.1109/TMTT.2014.2366108

Shi, J.-J., Chen, M.-S., & Wu, X.-L. (2010). A design of Ku-band slow-wave bandpass filter. *International Conference on Microwave and Millimeter Wave Technology*, 2010, 2063–2066. https://doi.org/10.1109/ICMMT.2010.5525210

Tseng, C.-H., & Chen, Y.-T. (2016). Design and implementation of New 3-dB Quadrature Couplers using PCB and silicon-based IPD Technologies. *IEEE Transactions on Components, Packaging and Manufacturing Technology*, 6(5), 675–682. https://doi.org/10.1109/TCPMT .2016.2550562

Varian, R. H., & Varian, S. F. (1939). A high frequency oscillator and amplifier. *Journal of Applied Physics*, 10(5), 321–327. https://doi.org/10.1063/1.1707311

Watkins, D. A. (1958). *Topics in Electromagnetic Theory*. Wiley.

Wu, K. (1999). Slow Wave Structures. In *Wiley Encyclopedia of Electrical and Electronics Engineering*. John Wiley & Sons, Inc. https://doi.org/10.1002/047134608X.W4952

Wu, K. (2005). Slow Wave Structures. In *Encyclopedia of RF and Microwave Engineering*. John Wiley & Sons, Inc. https://doi.org/10.1002/0471654507.eme402

Zhurbenko, V., Krozer, V., & Meincke, P. (2006). Miniature microwave bandpass filter based on EBG structures. *European Microwave Conference*, 2006, 792–794. https://doi.org/10.1109/ EUMC.2006.281038

2

Slow-Wave Coplanar Waveguides and Slow-Wave Coplanar Striplines

Anne-Laure Franc[1], Leonardo Gomes[2,3,4], Marc Margalef-Rovira[4], and Abdelhalim Saadi[5]

[1] *LAPLACE, Université de Toulouse, CNRS, INPT, UPS, Toulouse, France*
[2] *TIMA, Université Grenoble Alpes, Grenoble, France*
[3] *Polytechnic School, University of São Paulo, São Paulo, Brazil*
[4] *STMicroelectronics, RFC HDC, Crolles, France*
[5] *NXP, R&D HW RFP, Toulouse, France*

2.1 Introduction – Chapter Organization

As the integrated technologies become more complex, more and more metallic layers are available in the back-end-of-line. The conventional transmission lines (microstrips, coplanar waveguides (CPW), and coplanar striplines (CPS)) do not really take advantage of this flexibility. Various solutions have been implemented in different technologies so as to reach transmission lines with high slow-wave factors (SWFs): high-k dielectrics (Ma et al., 2008), reactive lumped elements loading (Bastida & Donzelli, 1979; Hirota et al., 1990) and distributed loading. The latter is very relevant because it can be implemented in conventional integrated technologies while providing many opportunities for the designer. Actually, the conductive layers of the back-end-of-line can be patterned in an advantageous way to overcome classical limitations. This is the case for the slow-wave coplanar waveguides (S-CPW), and slow-wave coplanar striplines (S-CPS), which are the building blocks of this chapter.

In Section 2.2, the S-CPW and S-CPS topologies are presented. They basically rely on a conventional CPW or CPS that is periodically loaded by an electrically floating shield, which consists of parallel ribbons placed orthogonally to the propagation direction. This conductive shield is designed to catch the electric field without significantly modifying the magnetic field distribution that flows through it. The separation of the magnetic and electric fields below the shielding layer explains the propagation of a slow-wave mode in the structure (Hasegawa et al., 1971), leading to a strong miniaturization in the longitudinal direction, as compared to conventional transmission lines.

The physical behavior being known, Sections 2.3 and 2.4, respectively, focus on the S-CPW and S-CPS, respectively. The analysis of the geometrical parameters highlights the high flexibility offered to the designer, and some design rules are presented to target efficient transmission lines depending on the desired application (for instance, high miniaturization or high-quality factor). The physical behavior of the slow-wave effect

Slow-Wave Microwave and mm-Wave Passive Circuits, First Edition. Edited by Philippe Ferrari,
Anne-Laure Franc, Marc Margalef-Rovira, Gustavo P. Rehder, and Ariana Lacorte Caniato Serrano.
© 2025 John Wiley & Sons Ltd. Published 2025 by John Wiley & Sons Ltd.

helps in understanding that the telegrapher's electrical model is not suitable for slow-wave transmission lines. Thus, another topology is given to electrically model slow-wave transmission lines, including the loss prediction.

Section 2.5 introduces coupled lines based on S-CPW. Here again, slow-wave principle in integrated technology introduces many degrees of freedom, especially a large electric and magnetic coupling range, thanks to an independent control of the even- and odd-mode propagation velocities. Various shielding plane arrangements (long full ribbons, cut ribbons at different places) can be imagined, and the electric model of each one is developed.

Finally, Section 2.6 presents many applications based on integrated slow-wave transmission lines, namely resonators, filters, power dividers, couplers, voltage-controlled oscillator (VCO) tanks, phase shifters, and sensors. This part aims at highlighting that any circuit based on transmission lines can benefit from the slow-wave effect to either improve its electrical performance, reduce its size, or add design flexibility. And most of the time, a bit of all three.

2.2 Principle of Slow-Wave CPW and Slow-Wave CPS

The coplanar transmission lines using slow-wave effect usually present the same general topology: A conductive patterned shield is placed below or above the coplanar conventional transmission line. This shield is composed of metallic ribbons orthogonally placed in the propagation direction. On the one hand, these ribbons catch the electric field and hence increase the electrical equivalent capacitance of the transmission line. On the other hand, they do not affect the magnetic field, and the electrical equivalent inductance remains almost unchanged. Thus, the phase velocity is slowed-down, leading to the slow-wave effect, as explained in Chapter 1.

If this principle can be implemented in various technologies as long as they present at least two metallic levels (one for the conventional coplanar transmission line and the second one for the shield), the most relevant is to use the shielding plane to mask the effect of a lossy substrate in addition to creating the slow-wave effect. For instance, when designed in complementary metal-oxide-semiconductor (CMOS) technologies, the patterned shield is advantageously placed below the transmission line so that the conventional transmission line benefits from the high thickness of the top metal level and the ribbons fully shield the silicon that introduces dielectric losses. That is why the following represents the shielding plane below the main transmission line, even if this is not mandatory to introduce slow-wave effect.

2.2.1 Slow-Wave Coplanar Waveguides Topology

As introduced in Chapter 1, the S-CPW is based on a standard CPW that experienced a distributed loading due to orthogonal periodic metallic ribbons (Fig. 2.1). The shield period has to be short enough compared to the wavelength to ensure that the loading is equivalent to a distributed phenomenon.

If the dimensions are correctly chosen (Franc et al., 2010), the shield catches the electric field, which does not penetrate through the low-resistivity silicon (for bulk technologies), hence increasing the capacitance while removing the silicon losses. This is verified

Figure 2.1 Slow-wave coplanar waveguide (S-CPW) and the associated electromagnetic field lines.

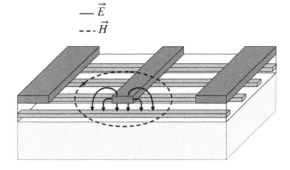

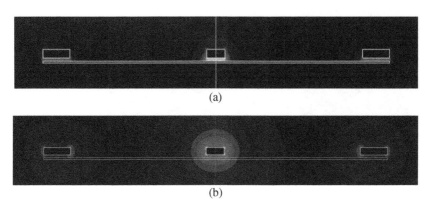

(a)

(b)

Figure 2.2 (a) Electric and (b) magnetic field magnitude in a S-CPW at 60 GHz, cut plane transversal to the propagation direction, based on classical S-CPW dimensions.

in Fig. 2.2(a), where the electrical field carried out with Flux3D (Cedrat, n.d.) was drawn for a typical S-CPW. At the same time, the magnetic field remains unchanged, as does the inductance (Franc et al., 2013b). Figure 2.2(b) shows the magnetic field magnitude of the same S-CPW: the magnetic field flows around and inside the floating ribbons the same way it does with a conventional CPW. Finally, the linear capacitance C_{lin} increases compared to a conventional CPW, and the linear inductance L_{lin} is almost the same, hence, from a circuit point of view, equation (2.1) shows that a slow-wave mode propagates in the S-CPW structure. In addition, as the characteristic impedance is computed in the lossless case with equation (2.2), the S-CPW topology allows maintaining a reasonable high value in comparison to microstrip or grounded-CPW, even if the high characteristic impedance that could be reached with CPW is no longer attainable.

$$v_\varphi = \frac{1}{\sqrt{L_{lin} \cdot C_{lin}}} \tag{2.1}$$

$$Z_c = \sqrt{\frac{L_{lin}}{C_{lin}}} \tag{2.2}$$

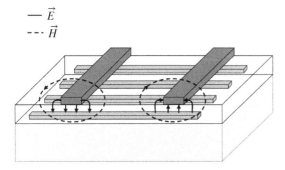

— $\vec{E}$
--- $\vec{H}$

Figure 2.3 Slow-wave coplanar stripline (S-CPS) and the associated electric and magnetic field lines.

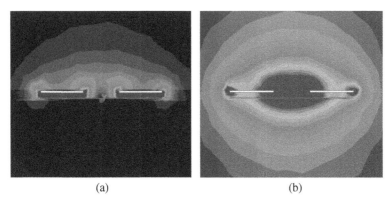

(a)   (b)

Figure 2.4 (a) Electric and (b) magnetic field magnitude in a S-CPS at 60 GHz, considering dimensions leading to an efficient shielding.

2.2.2 Slow-Wave Coplanar Striplines Topology

When differential transmission lines are required (for instance, to avoid the use of baluns), S-CPS are to be considered. The topology is shown in Fig. 2.3. It consists of a pair of parallel coplanar strip conductors that are loaded by a periodic plane composed of metallic ribbons orthogonally placed in the propagation direction, as for the S-CPW.

The principle is the same: the electric field is caught by the ribbons, while the magnetic field is hardly affected by the shield and flows through the ribbons. This is shown in Fig. 2.4 with full-wave simulations carried out with Ansys HFSS. Then, the equivalent linear inductance is the same, and the equivalent linear capacitance is increased compared to the unshielded coplanar stripline having the same dimensions. Eventually, the phase velocity decreases, and the transmission line exhibits a slow-wave effect.

2.2.3 Figures of Merit

The secondary parameters (the characteristic impedance Z_c, and the propagation constant γ) are extensively used to characterize transmission lines since they are key design parameters. The most common way to derive those electrical properties over a very wide frequency band is to use dedicated electromagnetic (EM) tools (computer-aided design (CAD) tools) that directly give S parameters.

Being aware of all the symmetries, the structure of interest is drawn and should then be properly excited by feeding with the correct electrical field. Depending on the chosen simulation tool and the case of the study, different ways of excitation are available and well documented for all classical transmission lines. After the simulation, this kind of CAD tool returns the S-parameters of the whole structure, including the contribution of the feeding sections between each port and the area of interest where the main propagating mode is established. Thus, a de-embedding step is needed to remove the feeding section contribution. One method consists of using two transmission lines with the same physical parameters except their lengths that vary (Mangan et al., 2006). This simple method was derived from the well-known Thru Reflect Line (TRL) calibration method (Engen & Hoer, 1979). The two lengths being l_1 and l_2, an equivalent transmission line having the same parameters with a length $|l_1 - l_2|$ is obtained after the de-embedding. In theory, the electrical characteristics of the equivalent transmission line should be independent of the values l_1 and l_2. In practice, some preliminary tests are required to obtain a good convergence. Note that this method, as derived from the TRL, is also relevant to remove the parasitics due to the pads and the feeding lines in the measurement steps.

Then, classical transmission line theory gives the equations (2.3) and (2.4) to extract the characteristic impedance Z_c and the propagation constant γ from the chain matrix $\begin{bmatrix} A & B \\ C & D \end{bmatrix}$. The propagation constant being known, the attenuation constant α and the phase constant β are easily extracted with $\gamma = \alpha + j\beta$. Then, it is possible to extract the quality factor of the transmission line and to qualify the slow-wave effect as defined in Chapter 1.

$$Z_c = \sqrt{\frac{B}{C}} \tag{2.3}$$

$$\gamma = \frac{\mathrm{Arccosh}(A)}{|l_1 - l_2|} \tag{2.4}$$

2.3 Slow-Wave Coplanar Waveguides

The S-CPW topology is recalled in Fig. 2.5. It consists of a conventional CPW whose strips thickness is t_{CPW}. The center signal strip has a width W, the ground strips have a width W_g and are separated from the center strip by the gap G. The introduction of the electrically floating shield adds many geometrical dimensions, namely the height h separating the two conductive layers, the thickness t_{shield} of the patterned shield, the ribbon width RW, and the ribbon spacing RS. Then, a designer has many degrees of freedom to implement a transmission line meeting the circuits' expectations.

2.3.1 Electrical Performance

Thereafter, various S-CPWs are studied in order to give some examples, clearly detail the effect of each geometrical parameter, and thus derive design rules. The main properties of the transmission lines are listed in Table 2.1; the back-end-of-line for each concerned technology is displayed in Fig. 2.6; and some pictures are provided in Fig. 2.7.

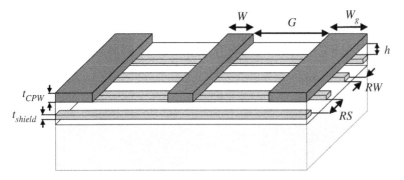

Figure 2.5 Geometrical dimensions of the S-CPW.

Table 2.1 Dimensions and characteristics of the S-CPWs considered in this chapter.

| | | Geometrical dimensions (µm) | | | | | | Electric performance at 60 GHz | | | |
| | | Main transmission line | | | Patterned shield | | | Z_c | | α | |
	Technology	W	G	W_g	RW	RS	h	(Ω)	ε_{reff}	(dB/mm)	Q
CPW_AMS		10	7	20	N/A	N/A	N/A	53	4.6	1.16	10
S-CPW1		10	100	60				40	35	1.07	31
S-CPW2	AMS 0.35 µm	18		60	0.6	0.6		31	47	1.17	33
S-CPW3			150				1	36	48	1.41	29
S-CPW4		7	50	10	0.7	0.7		53	24	0.99	27
S-CPW5					0.6	1		55	21.2	0.74	30
µstrip_B9MW		12.2	N/A	N/A	N/A	N/A	8.3	51	3.7	0.5	20
S-CPW6	STM B9MW 0.13 µm	5	50	10	0.16	0.64	0.4	35	25.6	0.66	43
S-CPW7							1.15	49	13.3	0.49	42
S-CPW8			40		0.2	0.8	3.05	70	8.2	0.48	38
µstrip_65 nm	STM 65 nm	3.8	N/A	N/A	N/A	N/A	3.62	47	3.45	1.55	6.5
S-CPW9		20	25	12	0.1	0.55	2.1	45	12.3	0.8	24

2.3.1.1 CPW Strips Dimensions

Thanks to many geometrical dimensions' choices, the designer may take advantage of several degrees of freedom available in S-CPW design. Hence, the signal width W, the signal-to-ground gaps G, and the oxide thickness h between the CPW and the shield, strongly affect the characteristic impedance and the phase constant. Typically, W is adjusted to reach the targeted characteristic impedance. The gap should be high enough – around some tens of micrometers – to ensure a reasonable high inductance and hence, a high SWF. The gap is also a lever to adjust the characteristic impedance value. For instance, the measured characteristic impedance and relative effective dielectric constant of three different transmission lines are plotted in Fig. 2.8, while the metal layers and the shield dimensions are

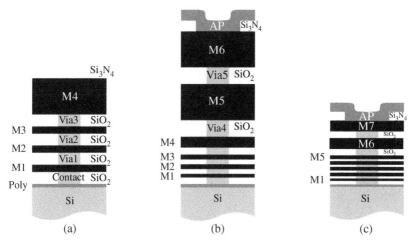

Figure 2.6 Back-end-of-line of the (a) AMS 0.35 µm; (b) STM BiCMOS9MW 0.13 µm; and (c) STM CMOS 65 nm.

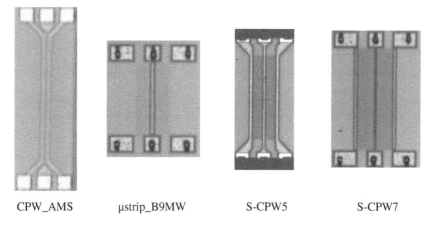

CPW_AMS   µstrip_B9MW   S-CPW5   S-CPW7

Figure 2.7 Photographs of some transmission lines.

the same. When the signal width increases, the capacitance also increases, and the characteristic impedance decreases. Similarly, an increase in the signal-to-ground gap leads to an increase in the inductance, leading to an increase in the characteristic impedance. Note that ripples in the characteristic impedance are due to the resonances of the measured transmission lines, which are electrically long thanks to the slow-wave effect.

When the designer aims for a strong slow-wave effect, the patterned shield is supposed to be not too far from the CPW strips (typically 1–5 µm). Nevertheless, one should check that the oxide remains thick enough to prevent electrical leakage through the patterned shield, as illustrated in Fig. 2.9. As a rule of thumb, h should be roughly kept higher than the ribbon spacing RS.

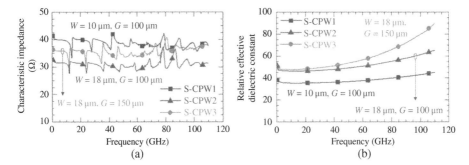

Figure 2.8 Measured (a) characteristic impedance and (b) relative effective dielectric constant for three S-CPWs in the same CMOS technology.

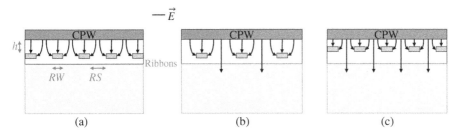

Figure 2.9 Electric field lines in a S-CPW for various *RS/h* ratios (cut plane in the propagation direction). (a)The electric field is totally captured by the floating shield; (b) the electric field is partially captured by the floating shield; and (c) for the same shielding ribbons dimensions as (a); the electric field is only partially captured by the floating shield because the dielectric thickness *h* is decreased.

Due to large signal-to-ground gaps, the S-CPW is quite large, so it is necessary to optimize the ground planes' widths. As most of the electrical field is concentrated in the oxide between the two conductor layers, the ground strips' widths of S-CPW can be fixed independently of the gap, contrary to classical CPW (in which the electrical field takes up most of the space between the signal strip and the ground strips). Actually, the ground strips' width can be reduced to very small values, but then the insertion loss may increase dramatically for very narrow ground strips. Besides, the slow-wave effect would be affected because the capacitance would be reduced, leading to a decrease in the relative effective dielectric constant (Fig. 2.10). In practice, a ground width W_g of roughly 10–15 μm is usually a good trade-off.

2.3.1.2 Shield Dimensions

Concerning the patterned shield, the period $RS + RW$ is set to a few microns (even shorter than 1 μm in the most advanced CMOS/Bipolar CMOS (BiCMOS) technologies) in order to ensure a distributed loading but also to minimize eddy current losses that would occur at high frequency if wide ribbons are used. The ribbon spacing RS has to be short enough to prevent any leakage from the electrical field, as already illustrated in Fig. 2.9 and discussed in Franc et al. (2010). At the same time, if the ribbon width RW is large, eddy currents develop easily in the metal, generating losses (Franc et al., 2013b). So RW should be reduced,

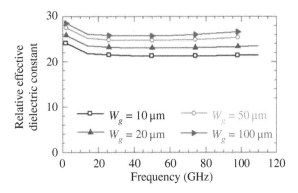

Figure 2.10 Simulated relative effective dielectric constant for various ground strips' widths.

but keep in mind that very thin parallel ribbons may introduce conductive losses too. The trade-off is discussed in Section 2.3.2, dealing with the electrical model of S-CPWs. The duty cycle defined as (2.5) is typically around 30% in the following practical case (Section 2.3.2.3) to ensure a minimum contribution of the floating shield to the total attenuation losses of the S-CPW. But its value also depends on the chosen technology and geometrical dimensions, which should then be studied for each new design. Moreover, the overlap between the ground strips and the patterned shield has to be large enough to ensure a high capacitive effect. So, the parallel ribbons are usually designed as long as the total S-CPW width (Tang et al., 2012).

$$d = \frac{RW}{RW + RS} \tag{2.5}$$

2.3.1.3 Metal Strips' Thickness
In CMOS technology, the CPW strips are usually implemented on one of the top metal layers. Indeed, the reduction of conductive losses in the CPW strips requires a thick metal, and the thickest ones are on top of the back-end-of-line. A thickness of 1 µm is usually enough to obtain acceptable losses. Nevertheless, 2 or 3-µm thick CPW strips are better suited, even at mm-waves. The skin depth δ, calculated with a very simplified model (2.6), is equal to 270 µm at $f = 60$ GHz for copper, μ_0 being the permeability in vacuum and σ the conductivity of the considered metal.

$$\delta = \frac{1}{\sqrt{\pi \, \mu_0 \, \sigma f}} \tag{2.6}$$

Once the CPW strips metal layer is determined, the patterned shield metal layer choice directly impacts the oxide thickness h and thus the capacitance value. If a metal near to the CPW strips is selected, say between 0.4 and 1 µm, a high slow-wave effect is achieved, and the characteristic impedance is likely to be low. For instance, the dielectric thickness equals 0.4 µm for the measurement results of the S-CPW6 shown in Fig. 2.11, the characteristic impedance is reasonably low (35 Ω), and the relative effective dielectric constant reaches 25.6, which means a slow-wave factor *swf* (as defined in Chapter 1) equal to 5.1, or a slow-wave factor *SWF* equal to 2.6 compared to a classical CPW or a microstrip line having a relative effective dielectric constant typically lower than 4. In the same manner, if a bottom metal layer is chosen for the floating shield ribbons, high characteristic impedances (~80 Ω) are easily reached at the expense of low slow-wave factors. As an illustration, the S-CPW8 in

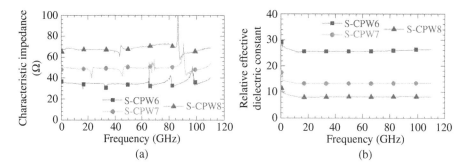

Figure 2.11 Measured (a) characteristic impedance and (b) relative effective dielectric constant for three different S-CPWs having various oxide thickness and gap dimensions.

Fig. 2.11 presents a 70-Ω characteristic impedance with a relative effective dielectric constant reduced to 8.2, i.e. an *swf* reduced to 2.8 or an *SWF* of 1.5. Once again, the design freedom degrees require determining a trade-off that depends on the targeted performance.

The S-CPW topology is optimal for the design of low characteristic impedance transmission lines (basically down to 15 Ω) with high slow-wave factors. It is more challenging to reach high characteristic impedance S-CPWs. Nevertheless, in many technologies, it is usually possible to reach 80 Ω, and sometimes 100 Ω, while maintaining a quality factor always higher than classical transmission lines, i.e. microstrip lines or CPWs. High characteristic impedance is an issue when dealing with conventional microstrip lines. Values higher than 75 Ω are basically not possible due to process limitations (minimum width limitations). Hence, the use of S-CPWs can be a solution to skirt this problem.

Whatever the targeted characteristic impedance, the relative effective dielectric constant of a S-CPW, and hence its slow-wave factor, is always higher than the one of a classical CPW or microstrip line. Remarkably high values (up to 60) may be reached by associating a high capacitance and a high inductance. The price to pay is either a large width of the S-CPW (gap increased to enhance the inductance) or a lower quality factor when the dielectric thickness is decreased in order to enhance the capacitance. As a rule of thumb, values around 20–25 can be achieved in quite all current technologies with high-quality factors (greater than 30) and moderate surface area. When wide gaps are considered to have high slow-wave factors, it is important to check that the final surface area is still competitive compared to classical solutions. Moreover, a large total width introduces interconnection problems, which will be addressed in the Section 2.6.1 dealing with slow-wave junctions.

Finally, high-quality factors (roughly up to 40–45 at 60 GHz) can be reached regardless of the targeted characteristic impedance, from 35 up to 70 Ω, see Fig. 2.12.

2.3.2 Electrical Model

The derivation of an equivalent electrical model is of high importance for designers as it drastically reduces the time-consuming full-wave electromagnetic simulations. The electrical model must be parameterized with all the geometric dimensions that constitute the circuit. Thanks to the many geometrical degrees of freedom existing in S-CPWs, a single design goal can be achieved through different architectures.

Figure 2.12 Quality factor for S-CPWs exhibiting characteristic impedances ranging from 35 to 70 Ω, in STM B9MW technology.

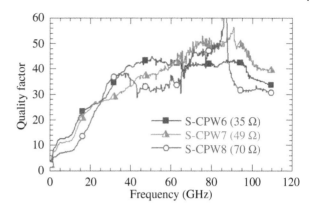

In this section, it is shown that the well-known telegrapher's *RLCG* model is not suitable when dealing with S-CPWs. A new topology is then proposed, and the way to derive a fully-parametrical electrical model with parameters only dependent on the geometry and material properties is presented.

2.3.2.1 Model Components

The classical telegrapher's model is a lumped-element representation of an infinitesimal length dx of a transmission line. This model allows to represent the distributed behavior of transmission lines using lumped elements while accurately describing the physical effects occurring in this kind of device. This model, displayed in Fig. 2.13, consists of R, L, C, and G elements: a series inductance L (in H/m) modeling the magnetic field effect; a series resistor R (in Ω/m) modeling metallic losses; a parallel capacitance C (in F/m) representing the electric field; and a parallel conductance G (in S/m) taking the dielectric losses into account.

As shown previously, the magnetic field has the same spatial distribution in a S-CPW, and in a classical CPW, the series inductance L remains valid and is almost equal for both topologies having the same geometrical dimensions. Similarly, the capacitance C is still present in the model. If the same lateral dimensions are considered, the capacitance per unit length of a S-CPW with its ribbons placed at a distance h to its signal and ground strips is almost equal to that of a fully grounded CPW with a lower ground placed at a distance h to its signal and ground strips. This can easily be explained by the fact that both the ribbons or the lower ground of the grounded CPW effectively capture the electric field, leading to very similar electrical field distributions for both architectures. The precise calculation of these components is given in Section 2.3.2.2. Nevertheless, the *RLCG* model is not adequate to evaluate the full behavior of S-CPW when losses are considered. Losses do not only come

Figure 2.13 Telegrapher's model.

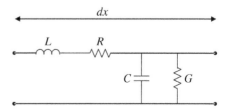

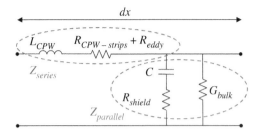

Figure 2.14 S-CPW electrical model.

from conductive losses in the CPW strips and substrate losses in the silicon. Conductive and eddy current losses also appear in the patterned metallic shield. For this reason, the model presented in Fig. 2.14 better represents the physical behavior of S-CPWs.

The resistance $R_{CPW-strips} + R_{eddy}$ includes conductive losses in the CPW strips and eddy current losses introduced by the patterned shield. Due to the capacitance between the CPW strips and the shield, a current propagates in the floating ribbons, leading to conductive losses in the shield modeled as a resistance R_{shield}. In addition, the conductance G_{bulk} expresses the losses in the bulk silicon: both conductive and eddy current losses. In (Tang et al., 2012), an additional study pointing out the effect of the floating ribbon width on the loss distribution for both standard Bulk and High Resistivity Substrate On Insulator (HR-SOI) silicon substrates demonstrated that eddy current losses induced by the magnetic field into the substrate are negligible for these transmission lines. Hence, in practice, if the shield is efficient enough to ensure that no electrical field couples to the substrate, this term can be neglected. The parameters can be extracted thanks to a quasi-static tool or empirical equations if the eddy currents are not considered, which is acceptable up to more than 100 GHz for most of the practical cases. Then, based on (2.7) and (2.8), R_{shield} and $R_{CPW-strips} + R_{eddy}$ can be calculated.

$$Z_{series} = jL\omega + R_{CPW-strips} + R_{eddy} \tag{2.7}$$

$$Z_{parallel} = R_{shield} + \frac{1}{jC\omega} \tag{2.8}$$

To accurately extract $R_{CPW-strips}$ and R_{eddy}, the designer can rely on an electromagnetic simulator allowing to compute the magnetic losses in the different volumes. This is possible when very short sections of the transmission line (for instance, only one period) are considered in a magnetic simulation. So, the term $R_{CPW-strips}$ can be extracted by observing the losses in the three CPW strips, independently from R_{eddy}.

The propagation parameters, i.e. characteristic impedance Z_c and propagation constant γ, are extracted with the classical equations (2.9) and (2.10), to give (2.11) and (2.12)

$$Z_c = \sqrt{Z_{series} \cdot Z_{parallel}} \tag{2.9}$$

$$\gamma = \sqrt{Z_{series}/Z_{parallel}} \tag{2.10}$$

$$Z_c = \sqrt{(jL\omega + R_{CPW-strips} + R_{eddy}) \cdot \left(R_{shield} + \frac{1}{jC\omega}\right)} \tag{2.11}$$

$$\gamma = \frac{j\omega\sqrt{LC}}{\sqrt{1 + (R_{shield}C\omega)^2}}\sqrt{\left(1 + \frac{R_{CPW-strips} + R_{eddy}}{jL\omega}\right) \cdot (1 - jR_{shield}C\omega)} \tag{2.12}$$

As $(R_{shield}C\omega)^2 \ll 1$ in practical cases, the propagation constant can be simplified (2.13). Then the phase constant and attenuation constant are finally expressed from the propagation constants, (2.14) and (2.15).

$$\gamma \approx j\omega\sqrt{LC} \cdot \left[1 - j \cdot \frac{1}{2} \cdot \left(\frac{R_{CPW-strips} + R_{eddy}}{L\omega} + R_{shield}C\omega\right)\right] \tag{2.13}$$

$$\beta \approx \omega \cdot \sqrt{L \cdot C} \tag{2.14}$$

$$\alpha \approx \frac{1}{2} \cdot \frac{R_{CPW-strips} + R_{Eddy}}{\sqrt{L/C}} + \frac{1}{2}\omega^2 \cdot R_{shield} \cdot C \cdot \sqrt{L \cdot C} \tag{2.15}$$

2.3.2.2 Model Component Calculations

Thanks to the fact that the behavior of the S-CPWs can be easily described using an ad-hoc version of the telegrapher's model (as for the other structures presented in this chapter), finding analytical ways to relate the geometrical dimensions of these transmission lines to the values of the model components is of crucial importance. Indeed, this allows the designer to have much more time-effective design sequences.

Resistance Computation In integrated technologies, the cross-section longitudes of metallic strips in the context of S-CPWs are often below 10 μm, especially for the height, which is in the order of magnitude of few micrometers. Hence, the skin effect can often be neglected. If that is not the case, the authors in Antonini et al. (1999) give a very accurate approach to modeling the resistance of rectangular – and not only – conductors.

In this context, the resistance per unit length of the signal/ground strips can be simply described as:

$$R_{CPW} = \delta x \cdot \frac{\rho}{W \cdot t_{CPW}} \tag{2.16}$$

where δx is the unit length considered in the model, ρ is the resistivity of the metal, W represents the width of the strip, and t_{CPW} stands for the thickness of the metal used to integrate the strips.

In the first approximation, the resistance of the metallic ribbons is usually negligible. However, if it is necessary for the convergence of the model (i.e. ribbons being very lossy or very densely distributed), one can simply use equation (2.16) to obtain their resistance.

Note that the eddy current effect is not considered in this calculation.

Capacitance Computation As compared to a classical CPW electrical model, the capacitance computation can differ. An innovative and yet extremely powerful approach to calculate capacitances in the context of S-CPWs was proposed in Bautista et al. (2015). In order to calculate the capacitance, the electric field around the conductors is separated in four regions, namely the bottom plate, the point charge, the fringe, and the upper-plate capacitances, as shown in Fig. 2.15. The total capacitance C_{Tot} is equal to the sum of the four elementary capacitances:

$$C_{Tot} = C_{plate} + C_{point-charge} + C_{fringe} + C_{up} \tag{2.17}$$

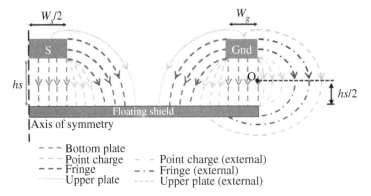

Figure 2.15 S-CPW electric field lines.

First, let's describe the well-known equation of the parallel plate capacitance:

$$C_{plate} = \frac{\varepsilon_0 \cdot \varepsilon_r \cdot W \cdot \delta x}{h_s} \tag{2.18}$$

where ε_0 and ε_r represent the vacuum dielectric constant and the relative dielectric constant of the material where the electric field is confined, respectively.

Next, the point-charge capacitance can be described as follows:

$$C_{point-charge} \approx 0.82 \cdot \varepsilon_0 \cdot \varepsilon_r \cdot \delta x \tag{2.19}$$

On the other hand, the fringe capacitance can be described in the following form:

$$C_{fringe} = \frac{2 \cdot \varepsilon_0 \cdot \varepsilon_r \cdot t_{CPW} \cdot \delta x}{\pi} \left(\sum_{i=1}^{n} \frac{1}{(i-1) \cdot t_{CPW} + n \cdot h_s} \right) \tag{2.20}$$

where n is an empirical parameter allowing to tune the granularity of the calculation. The greater n is, the more precise the calculation will be, but the computational time will also be increased.

Finally, the up-capacitance can be computed using the following equation:

$$C_{up} = \frac{2 \cdot \varepsilon_0 \cdot \varepsilon_r \cdot W \cdot \delta x}{n} \left(\sum_{i=1}^{n} \frac{1}{ri_{up} \cdot \left[\pi + 2\cos^{-1}\left(\frac{h_s + t_{CPW}}{ri_{up}} \right) \right]} \right) \tag{2.21}$$

where ri_{up} is the radius described by the electrical field:

$$ri_{up} = (h_s + t_{CPW}) \cdot \left(1 - \frac{i-1}{n} \right) + \left(\frac{i-1}{n} \right) \cdot h_{max} \tag{2.22}$$

and h_{max} is the maximum height achieved by these electrical lines:

$$h_{max} = \sqrt{(h_s + t_{CPW})^2 + \left(\frac{W}{2} \right)^2} \tag{2.23}$$

Note that here all field lines are considered to go from the signal/ground strips to the metallic shield. In the case of narrow gaps between the signal and ground strips, the up capacitance, the fringe, point charge, or even the plate capacitance can have shorter paths to other strips than to the metallic shield. To properly describe the behavior of these geometries, the path leading to a metallic surface that leads to a shorter distance must be considered. This proves especially important in the context of coupled S-CPW, where the distance between the signal strips tends to be reduced. This case is partially discussed in Bautista et al. (2015) and further studied in Margalef-Rovira (2020).

Inductance Computation   As previously introduced, the use of a metallic shield does not modify the magnetic field behavior. If the frequency is sufficiently low to neglect the skin effect, equations (8) and (9) from Zhong and Koh (2003) can be used to calculate the quasi-static partial self- and mutual-inductances between the signal and ground strips and even the ribbons. This is the case for most practical cases.

However, there are two cases where this approach may not be sufficient: (i) a very high frequency where the skin effect is no longer negligible or (ii) structures where the signal/ground or signal-signal distance is reduced (as for couplers), leading to large magnetic coupling. In these cases, the designer may have to use more advanced methods where the distribution of the current within the strips is considered. The work in Margalef-Rovira (2020) introduces a solution to this issue in the context of coupled S-CPW that can be applied to other propagating structures without loss of generality. This work is built on the methods proposed by Paul (2003).

2.3.2.3 Losses Distribution

The three losses contributions are clearly apparent in (2.15). The first two terms are related to the ohmic losses in the CPW strips and to the eddy currents flowing inside the shielding ribbons. The last one corresponds to the conductive losses in the shielding ribbons and increases very quickly with both the frequency and the capacitance. The designer should keep in mind that a high slow-wave factor (which means a high capacitance and thus low characteristic impedance) leads to higher losses due to a significant increase in the shield conductive losses. The conductive losses in the shield are directly linked to the value of R_{shield} which is calculated as:

$$R_{shield} = \frac{RW + RS}{2} \cdot \frac{W/2 + G + W_g}{\sigma_{shield} \cdot RW \cdot t_{shield}} = \frac{1}{2 \cdot d} \cdot \frac{W/2 + G + W_g}{\sigma_{shield} \cdot t_{shield}} \qquad (2.24)$$

t_{shield} being the patterned shield thickness and σ_{shield} its conductivity. R_{shield} can be decreased thanks to an increase either in the duty cycle d or in the shield thickness t_{shield}. In both cases, the eddy current loops may more easily develop in the bigger metallic patterned ribbons volume. So generally speaking, a set of optimal shield dimensions exists to minimize the total losses. This optimal choice depends on all the geometrical dimensions of the transmission line. As an example, the case of S-CPW5 is studied in detail thereafter.

Figure 2.16 shows the loss distribution when the shielding ribbons thickness varies. Quite logically, it appears that the losses due to the CPW strips are not affected by the increase in shielding ribbons thickness. In this particular case, the eddy current losses are negligible, but their proportion increases with thicker shielding ribbons, whereas the conductive losses proportion decreases.

In Fig. 2.17, the conduction losses and the eddy current losses in the shielding ribbons are studied versus frequency for various duty cycle values, with the shield period $RW + RS$ kept

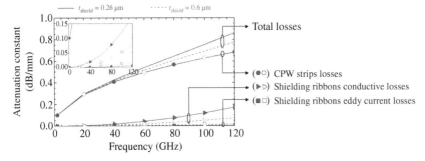

Figure 2.16 Simulated attenuation constant and losses distribution for two shielding ribbons thicknesses, based on S-CPW5 dimensions. Inset: zoom for low attenuation constant values.

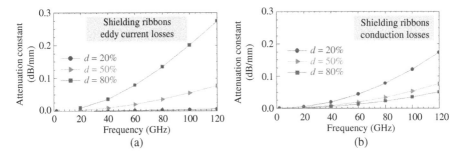

Figure 2.17 Simulated losses in the floating ribbons for various duty cycles: (a) eddy current losses and (b) conduction losses. (S-CPW5 dimensions were considered.)

constant and equal to 0.8 μm. As already mentioned, the eddy current losses increase very fast with the duty cycle, while the shielding ribbon conduction losses decrease. For instance, at 120 GHz, the eddy current losses represent less than 1% of the total attenuation constant for a duty cycle of 20%, whereas they represent at least one-quarter of the total losses for a duty cycle of 80%. At the same time, the conductive losses in the floating ribbons decrease from 20 to 5% of the total attenuation constant. Finally, in this specific worst case (duty cycle of 80% and working frequency of 120 GHz), the losses generated in the shielding ribbons represent almost one-third of the total losses shown in Fig. 2.16.

Then, considering the total amount of losses, a trade-off for these particular dimensions can be deduced from the plots in Fig. 2.18. As the CPW strip losses contribution does not vary much with the floating shield duty cycle, the best trade-off is obtained when the sum of the losses in the floating ribbons, i.e. conduction losses and eddy current losses, is the lowest. This trade-off corresponds to a duty cycle around 30–35% at both 60 and 120 GHz for the considered work case.

As shown previously, the main part of the attenuation constant comes from the CPW strips. Besides the shielding ribbon losses (conduction and eddy current losses), depend on the characteristic impedance value. When it is high enough, i.e. for low capacitance and low slow-wave effect, there is no need for strong optimization for the shielding ribbons dimensions. But when the capacitance is high, the optimization of the duty cycle appears to be of crucial interest. Then, thin and narrow parallel floating ribbons should be preferred,

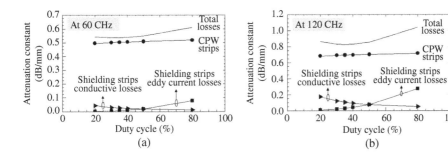

Figure 2.18 Simulated attenuation constant repartition versus the duty cycle at (a) 60 GHz and (b) 120 GHz. (S-CPW5 dimensions were considered.)

keeping in mind that (i) the slow-wave effect requires all the electric field to be confined between the CPW strips and the shielding layer, and (ii) the advanced technologies allow such thin metallic layers that the conductive losses in the parallel ribbons may become significant.

2.3.2.4 Dispersion: Floating Shield Equivalent Inductance

As already mentioned, the phase constant β is almost the same as that derived from a classical RLCG model. However, if one observes the measurement results in Fig. 2.8(b), it is obvious that dispersion occurs. This is due to the very wide gaps of these S-CPWs, i.e. 100 μm for S-CPW1 and S-CPW2, and 150 μm for S-CPW3, respectively. This dispersion cannot be explained by the electrical model proposed in Fig. 2.14. In order to take this dispersion into account in the electrical model, it is necessary to add an inductance L_{shield} in series with the capacitance. This inductance reflects the propagation along the floating ribbons. It was not initially introduced in the model presented in Fig. 2.14 because it only concerns S-CPWs carried out with very wide gaps, which is not the trend in current and future CMOS or BiCMOS technologies, because of the mm² cost.

Figure 2.19 gives the complete equivalent electrical model, also valid for S-CPWs with wide gaps. However, it is worth mentioning that L_{shield} should be considered for the S-CPWs with narrow gaps when they are used at very high frequencies; roughly above few hundreds of GHz. In that case, the impedance $L_{shield} \cdot \omega$ will no longer be negligible.

The same calculation as the previous one can be derived. The easiest way to obtain the final result is to consider an equivalent frequency-dependent capacitance $C'(f)$ as expressed in (2.25). The phase constant is now frequency-dependent and increases with frequency, accounting for dispersive wide-gap S-CPW. This phenomenon is confirmed by the plots in Fig. 2.20(a) presenting the relative effective dielectric constant for different S-CPWs with various total widths. S-CPW4, having a narrower gap equal to only 50 μm, does not present

Figure 2.19 S-CPW equivalent electrical model with floating ribbons equivalent inductance.

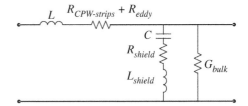

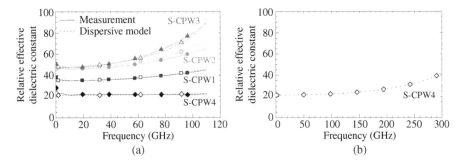

Figure 2.20 Relative effective dielectric constant of S-CPWs having various gap widths, in the AMS 0.35 µm technology. (a) Comparison of measurements and calculus; and (b) calculus up to 300 GHz for the narrower S-CPW.

dispersion up to 110 GHz. Figure 2.20(b) shows the calculated relative effective dielectric constant for S-CPW4 with a 50-µm gap up to 300 GHz. This confirms that dispersion also occurs for narrow gaps at high frequencies when the impedance $L_{shield} \cdot \omega$ is no longer negligible.

$$C'(f) = \frac{C}{1 + (j \cdot \omega)^2 \cdot L_{shield} \cdot C} \tag{2.25}$$

2.3.3 Benchmark With Conventional Transmission Lines

2.3.3.1 Comparison of Electrical Performance

Figure 2.21 compares the main electric characteristics of conventional and S-CPW transmission lines, whose dimensions are given in Table 2.1. Two different CMOS technologies and two kinds of conventional topologies (CPW and microstrip lines) are considered. The comparison was carried out by considering the same characteristic impedance of c. 50 Ω.

The relative effective dielectric constant of the conventional structures is basically limited by the intermetal dielectrics in the back-end-of-line (mostly SiO_2, whose relative dielectric constant is usually 3.9). It can be increased by leaving the passivation layer (mostly Si_3N_4, whose relative dielectric constant is roughly 7) above the strips. However, the S-CPWs relative effective dielectric constants are always much higher. The slow-wave factors *SWF* of S-CPWs typically range from 1.5 to more than 3, depending on the characteristic impedance, compared to those of classical transmission lines.

Concerning the losses, the attenuation constant strongly depends on the chosen technology and also on the dimensions. Nevertheless, for optimized structures using the same CMOS technology, the attenuation constant is in the same order of magnitude for classical and S-CPWs. Thus, the quality factors of S-CPWs are 1.5–3 times higher compared to those of classical transmission lines.

Figure 2.22 shows a representation to benchmark S-CPW against conventional topologies (CPWs and microstrip lines). Beside the AMS 0.35 µm, STM BiCMOS9MW 0.13 µm, and STM CMOS 65 nm from Table 2.1, the following transmission lines are displayed on the benchmark: microstrip lines extracted from Quémerais et al. (2010), Cathelin et al. (2007), and Morandini et al. (2010), CPWs from Cathelin et al. (2007), and Giancsello et al. (2007),

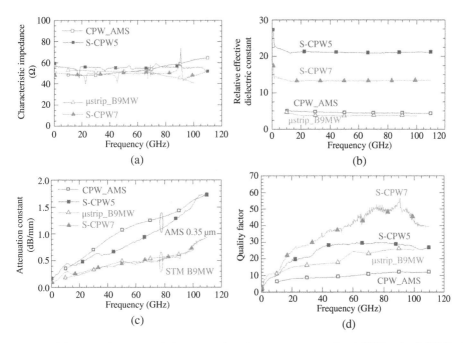

Figure 2.21 Electric characteristics comparison between a conventional CPW and a S-CPW in AMS 0.35 μm, and between a conventional microstrip transmission line and S-CPW in STM BiCMOS9MW 0.13 μm. (a) Characteristic impedance; (b) relative effective dielectric constant; (c) attenuation constant and (d) quality factor.

Figure 2.22 60-GHz measurement: Quality factor versus attenuation constant for transmission lines in various technologies.

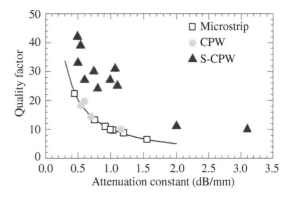

and S-CPWs from Kaddour et al. (2009), Franc et al. (2011), Cheung & Long (2006), Vecchi et al. (2009), Lai & Fujishima (2007), and Lee and Park (2010).

The measured quality factor is plotted versus the attenuation constant for a characteristic impedance around 50 Ω, and the benchmark is realized at 60 GHz. (Note that the characteristic impedance for Lai and Fujishima (2007) is unknown and that the results in Cheung and Long (2006) are given at 40 GHz.) This plot obviously exhibits a hyperbola for conventional topologies. Hence, the phase constant only depends on the intermetal relative dielectric constant and is roughly the same for all the considered CMOS technologies. Then, the hyperbola comes directly from the chosen quality factor definition. Through an increase in the phase constant, the slow-wave phenomenon frees the quality factor from the hyperbola, as shown in Fig. 2.22. So, for a given attenuation constant, much higher quality factors may be obtained, which also means a longitudinal miniaturization for a targeted electrical length.

At first sight, one thinks that the top left area in Fig. 2.22 is the most interesting area, because it corresponds to high-quality factor with low attenuation constant. On further consideration, the top right area offers better performance, because the same quality factor is obtained with highest attenuation constant, meaning that the highest relative effective dielectric constant is obtained, hence leading to more compact transmission lines. Note that to be comprehensive, one must consider not only the length but also the width of the transmission line; the footprint area is the parameter to take into account, as discussed in the next section.

In high-resistivity silicon on insulator (HR-SOI) technology, the comparison of S-CPW versus CPW can be done in order to demonstrate once again the interest of these slow-wave structures. Even if losses into the low-resistivity substrate are negligible, the attenuation losses of S-CPWs can be close to CPWs, whereas additional losses due to the shielding ribbons are expected for S-CPWs. Indeed, especially to achieve low characteristic impedance, a high capacitive effect must be achieved. For CPW, to obtain a reasonable lateral dimension of the structure, a smaller gap width G must be considered compared to S-CPW. Hence, higher losses linked to proximity effects between the ground planes and the signal strip are achieved for CPWs as compared to S-CPW (Tang et al., 2012).

2.3.3.2 Trade-off Between Surface Area and Electrical Performance

In comparison to conventional CPW or microstrip lines, the S-CPW topology gives more design degrees of freedom. So, several groups of geometrical dimensions give similar results in terms of characteristic impedance or electrical length. At the same time, wide S-CPWs are required to reach strong slow-wave effects. So, depending on the application, a trade-off has to be defined between the targeted electrical performance and the maximum silicon footprint. Fig. 2.23 helps in understanding this point. It represents the electrical length per mm^2 corresponding to the transmission lines described in Table 2.1 versus the relative effective dielectric constant. The surface area was calculated as the physical length multiplied by the total width ($W + 2G + 2W_g$). The highest relative effective dielectric constant does not give the lowest footprint area because a high relative effective dielectric constant is

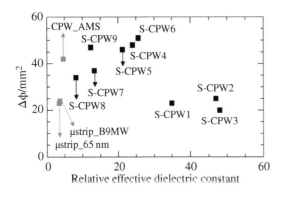

Figure 2.23 Electrical length per surface area versus the relative effective dielectric constant.

obtained thanks to wide gaps, leading to large widths. CPW and microstrip lines lead to poor performance, in terms of footprint area and quality factor. And finally, the S-CPW6 with a 50-Ω characteristic impedance leads to the best electrical performance, with a quality factor greater than 40 along with the smallest footprint area.

2.4 Slow-Wave Coplanar Striplines

The S-CPS topology is recalled in Fig. 2.24. This structure, is of course, very close to S-CPW, as CPSs are close to CPWs. The fundamental difference is related to the fact that it propagates only one mode, which is qualified as odd-mode since the structure is purely differential. Two metallic conductors of thicknesses t_{CPS} and t_{shield} are required. One consists of the conventional CPS with two parallel main strips of widths W_{s1} and W_{s2} and separated by a gap G. The second metal level is placed at a height h from the CPS strips, and it is patterned with electrically floating ribbons orthogonally placed in the propagation direction. Those ribbons have a width RW and a spacing RS. Their length is defined as $W_{s1} + W_{s2} + G + 2O$, where O is the overhang.

2.4.1 Electrical Performance

As the topology is similar to the S-CPW one, all the design rules associated with the dimensioning remain the same for the S-CPS. In the following, only the main results are given to highlight key trends, and specific attention is paid to the overhang, which was not defined previously. In addition, a special focus is placed on an interesting performance featured by S-CPS, namely the wide range of synthesizable characteristic impedances.

The variations of conventional CPS dimensions are reported in Suh and Chang (2002). When dealing with S-CPS, similar results are obtained: the characteristic impedance increases with the gap G, and the losses decrease for wider gaps. The relative effective dielectric constant increases with the gap G. So large gaps enhance the slow-wave effect, thus the longitudinal miniaturization, but it also increases the total width of the CPS, and consequently, a trade-off should be realized to miniaturize the total footprint of the S-CPS.

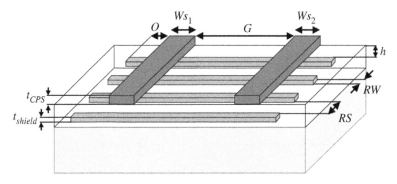

Figure 2.24 Geometrical dimensions of the S-CPS.

Concerning the shielding, the sizing is carried out in the same way as for S-CPWs. When the slow-wave effect is ensured, a short ribbon width RW is preferred, as it limits the attenuation loss (Abdel Aziz et al., 2009).

Once again, similar conclusions are obtained regarding the metal thicknesses if we consider S-CPSs or S-CPWs. As most of the total loss is due to conduction in the CPS strips, the CPS should be implemented in the thickest metal available. Regarding the patterned ribbons, a trade-off has to be defined depending on the working frequency because a thick metal layer would increase the eddy currents but a too thin shielding plane would increase the conductive losses, as for the S-CPWs.

The overhang O represents the overlap between each coplanar strip and the ribbons underneath: ribbons longer than the whole CPS's width ($W_{s1} + W_{s2} + G$) lead to a positive overhang, while a negative overhang is defined for ribbons that do not exceed the total CPS width. As the slow-wave effect is directly related to the capacitance between the coplanar strips and the shield, it strongly depends on the overhang value. The equivalent capacitance increases with the overhang, leading to a lower characteristic impedance and a higher slow-wave factor, Fig. 2.25. Then, the properties saturate, because increasing the length of the ribbons beyond the CPS width does not affect the electric field anymore. So, an overhang O of 0 (exactly the whole CPS's width) is an appropriate choice. The photograph of the associated S-CPS fed with ground–signal probes is shown in Fig. 2.26.

A set of S-CPSs was also considered in STM BiCMOS 55 nm. Their main properties and their 60-GHz performance are summarized in Table 2.2. The associated chip photographs

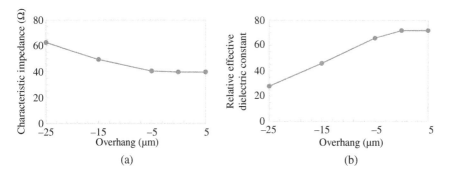

Figure 2.25 10-GHz simulated characteristic impedance (a) and relative effective dielectric constant (b) versus the overhang O. ($W_{s1} = W_{s2} = 35\,\mu m$, $G = 40\,\mu m$, $h = 1\,\mu m$, $RS = RW = 0.6\,\mu m$).

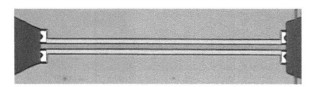

Figure 2.26 Photograph of a S-CPS in AMS 0.35 µm ($W_{s1} = W_{s2} = 35\,\mu m$, $G = 40\,\mu m$, $h = 1\,\mu m$, $RS = RW = 0.6\,\mu m$).

Table 2.2 Dimensions and characteristics of the S-CPSs considered in this section.

		Geometrical dimensions (μm)								Electric performance at 60 GHz			
		Main transmission line				Patterned shield				Z_c		α	
	Technology	W_{s1}	W_{s2}	G	t_{cps}	RW	RS	O	h	(Ω)	ε_{reff}	(dB/mm)	Q
S-CPS1		10	10	7.6	5.4				2	50	6.7	1.5	10
S-CPS2	BiCMOS	1	1	2.8	5.4	0.6	0.6	0	2	45	5.9	1.8	7
S-CPS3	55-nm	5	5	1.5	5.4				2	23	7.8	2.6	6
S-CPS4		5	5	20	3				4	105	5.3	1.3	10

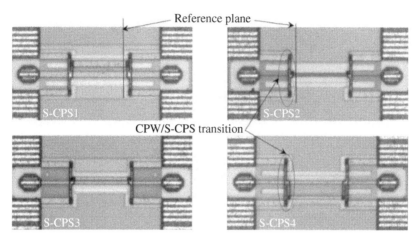

Figure 2.27 Photographs of S-CPSs.

are given in Fig. 2.27. Note that in these cases, the prober was equipped with ground–signal–ground probes, which are not suitable to carry out differential measurements. This constraint has been addressed by using a CPW to CPS transition, but it affects the quality factor, as shown in Fig. 2.28. Electromagnetic simulations carried out with Ansys HFSS show that similar quality factors as compared to S-CPWs should be reached.

The same shielding is used for the S-CPS1 and S-CPS2, which exhibit similar characteristic impedance but different slow-wave effects due to different CPS geometrical dimensions. The S-CPS3 and S-CPS4 highlight the very wide characteristic impedance range that can be achieved with S-CPSs. Hence, with the same strip widths, increasing the gap at the same time as increasing the height between the CPS strips and the shield allows to drastically increase the characteristic impedance.

2.4.2 Electrical Model

A classical RLCG model cannot be used, as for S-CPWs. The capacitive coupling to the shielding ribbons and the resistance of these ribbons are predominant in S-CPSs. Hence, the model described in Fig. 2.19 must be used to describe the physical performance of the

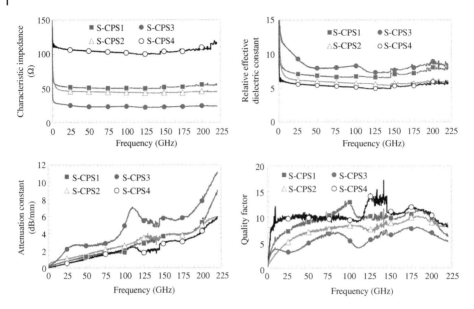

Figure 2.28 Measured data of the S-CPSs.

S-CPSs. As for S-CPWs, the inductance of the shielding ribbons has only been considered when the operating frequency or the lateral dimensions of these devices are large.

2.4.3 Design

2.4.3.1 Design Rules
The main advantages of slow-wave transmission lines (S-CPWs and S-CPSs) are their good electrical performance (higher quality factor than conventional transmission lines for similar losses) and the reduced size for a given length (the slow-wave factor enabling the miniaturization). This is obtained thanks to the implementation of a patterned shield plane that introduces many additional degrees of freedom: mainly the width RW and space RS of the parallel ribbons as well as the height h between the strips and the shielding ribbons. These degrees of freedom increase the design complexity; hence, some design rules are suggested to help the designer make the best choice with the various degrees of freedom.

Firstly, the oxide thickness can be fixed for a given technological stack-up. The choice of the oxide thickness h for a specific slow-wave transmission line should be done by considering a trade-off between the electrical performance, such as the attenuation constant and the slow-wave factor, and the range of attainable characteristic impedances. Those ranges can be obtained by varying the dimensions of the CPW or CPS strips when the slow-wave effect is properly realized. To ensure this last point, it is important to keep in mind that the slow-wave principle is based on the physical separation of the electric and magnetic fields. The ribbon spacing RS should be smaller or similar to the oxide height h to properly catch the electric field, but it cannot be too small because the magnetic field must flow through the patterned shield.

2.4.3.2 Design Flexibility

Some Possibilities in Integrated Technologies In the general case, two metallic layers are enough for the design of slow-wave transmission lines. Note that it can be interesting to use two oxide thicknesses for flexibility, specifically to enlarge the characteristic impedance range. Usually the back-end-of-lines offer thick top metal(s) that can be used for the strips and several thin metal layers below in which the floating ribbons can be patterned. This combination allows for minimizing the losses; nevertheless, many possibilities can be envisioned depending on the technology.

Optimization Based on Design Constraints Another good point of the slow-wave transmission lines is the high flexibility offered to the designer. Actually, a transmission line with the same characteristic impedance and electrical length can be obtained with different combinations of physical dimensions, leading to different slow-wave factors. As the miniaturization is not the same, the physical footprint is different. This is an interesting point because the designer has significant flexibility to take the systems' physical constraints into account, like in antenna array design or to align the input and output of a circuit.

2.5 Coupled Slow-Wave Coplanar Waveguides

High frequency couplers can serve a multitude of applications, such as power dividing/combining, I/Q generation, and phase shifting. Due to their versatility, they are often found in many electronic circuits. As for other transmission line topologies, couplers can be designed using S-CPWs, yielding the so-called coupled S-CPW structure.

2.5.1 Topology

Beyond the fact that S-CPWs spatially separate the electrical and magnetic fields, as compared to classical CPWs, no other major difference is observed between these two structures. Hence, the strategy adopted to design coupled S-CPWs is similar to the one used to integrate coupled CPWs, as shown in Fig. 2.29.

2.5.1.1 Design Flexibility

Transmission lines are roughly designed to match a certain impedance and electrical length. On the other hand, couplers usually require an accurate design in terms of electrical length,

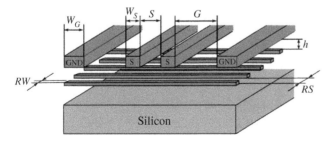

Figure 2.29 Generic coupled S-CPW architecture and its geometrical dimensions.

even- and odd-mode characteristic impedances, and coupling. Their design is thus more complex than their transmission line counterparts. In this scenario, having a floating shield below the signal strips adds an extra degree of flexibility to the design.

Center-Cut Ribbons Indeed, in a coupler-like structure, the ribbons do not only serve to modify the phase constant of the transmission line but also greatly impact the coupling between the two signal strips of the structure. Hence, adding modifications to it varies the coupling coefficient of the coupler.

This is the scope of the coupled S-CPWs with a center-cut architecture, with floating ribbons cut between the two signal strips, as shown in Fig. 2.30. As it is demonstrated subsequently in Section 2.5.4, the center-cut architecture provides an extra degree of freedom for the determination of the odd-mode characteristic impedance and the coupling coefficient.

In a first qualitative approximation, this can easily be seen if the reader pictures the fact that when a center-cut is introduced, the electrical coupling between the two strips is greatly reduced, and hence their odd-mode characteristics are modified. However, when an equal signal is present in both signal strips (i.e. even-mode), the introduced modification has no effect since the electrical coupling between two points with the same signal is null.

Side-Cut Ribbons A side-cut to the ribbons can also be performed in order to modify the coupled line characteristics, as shown in Fig. 2.31.

The side-cut architecture produces the opposite effect to the center-cut architecture. Indeed, when a side-cut is introduced, the odd-mode characteristic impedance is not modified, while the coupling coefficient and the even-mode characteristic impedance are, as described in Section 2.5.4. This can again be easily pictured if one notices that, with the addition of the side-cut, the electrical coupling to the ground strips is reduced, greatly affecting the even-mode propagation. On the other hand, the virtual ground still exists

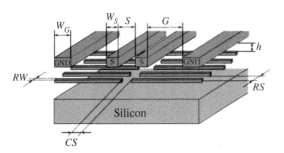

Figure 2.30 Generic coupled S-CPW with center-cut architecture.

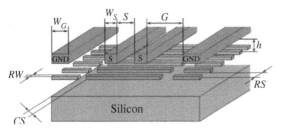

Figure 2.31 Generic coupled S-CPW with side-cut architecture.

in the middle of the structure for the odd-mode, which makes the effect of the side-cut negligible.

2.5.2 Electric and Magnetic Fields Distribution

Concerning the spatial distribution of the electromagnetic field, the coupled S-CPW structure is similar to a S-CPW. As shown in Fig. 2.32, which displays the electric and magnetic fields in a classical coupled S-CPW, the electric field is captured by the ribbons while the magnetic field flows around the conductors as in a classical coupled CPW. It is also important to point out that, as this structure contains two signal strips, electromagnetic coupling between these two exists and can be observed in Fig. 2.32.

Finally, due to the similar distribution of the electromagnetic field as compared to a S-CPW, similar considerations as the ones described in Section 2.3.1.1 can be considered by the designer in order to avoid undesired coupling with the substrate.

2.5.3 Propagation Modes in Coupled Slow-Wave CPWs

Due to the longitudinal symmetry observed in a coupled S-CPW, these structures can hold three modes of propagation: (i) even-mode, (ii) odd-mode, and (iii) mixed-mode. The two first modes can be excited by exciting two of the ports found in the same plane with

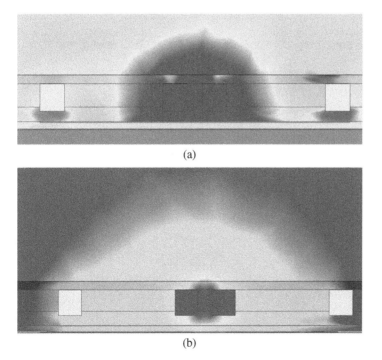

(a)

(b)

Figure 2.32 (a) Electric and (b) magnetic field magnitude in a coupled S-CPW at 60 GHz, cut plane transversal to the propagation direction, based on classical S-CPW dimensions.

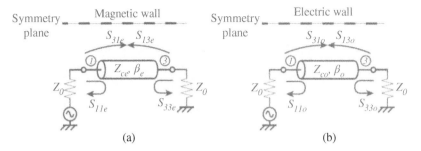

Figure 2.33 Modal analysis schematics of coupled lines: (a) even-mode and (b) odd-mode.

in- and complementary-phase signals, for the even- and odd-modes, respectively. Feeding the coupler with any other excitation results in mixed-modes.

When analyzing the even- and odd-modes, we can use some considerations that most currently commercial full-wave electromagnetic simulators integrate (i.e. placing a magnetic or electric wall following the symmetry plane), as shown in Fig. 2.33. The existence of a virtual magnetic or electric wall in the structure symmetry plane is the consequence of an even- or odd-mode excitation. In addition, this approach yields a more compact and simpler structure that can reduce the simulation time and the analysis complexity.

In the following section, the equivalent distributed RLRC model will be analyzed using the even- and odd-mode analysis. The mixed-mode behavior can be subsequently derived from these two propagation modes.

2.5.4 Definition of the Electric Model Topology: RLRC Model for Coupled Lines

As mentioned earlier, an additional degree of freedom was introduced with three different shielding configurations, namely: uncut, center-cut, and side-cut (Lugo-alvarez et al., 2014). In the center-cut topology, a cut is made at the center of the shielding ribbons. Conversely, in the side-cut shielding, cuts are made at the sides of the ribbons (refer to Fig. 2.34). In the uncut shielding, the floating ribbons remain intact. These configurations provide the designer with the ability to adjust the electric coupling coefficient of the structure, k_C, while keeping its magnetic coupling coefficient, k_L, unchanged.

Figure 2.34 presents the distributed model of the coupled S-CPW architecture in its three declinations. This model is derived from the work presented in Section 2.3.2.

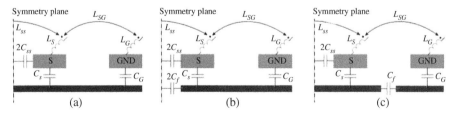

Figure 2.34 Distributed model elementary cell of a lossless coupled S-CPW with central structure symmetry with (a) uncut shield; (b) center-cut shield; and (c) side-cut shield

For simplicity, let's assume the coupled S-CPWs are lossless, and there is minimal coupling between the signal and ground strips. This assumption is valid in most coupler implementations since the distance between the strips and the floating shield, h, is typically smaller than the gap between the signal and ground strips, G. Additionally, we will assume that the gap G is less than $100\,\mu m$, at which point the propagation in the ribbons begins to exhibit dispersion in the effective relative dielectric constant of S-CPWs (see Section 2.3.2.4).

By utilizing these models, a modal analysis can be conducted to determine k_L and k_C, as well as the characteristic impedances of the even-mode (Z_{even}) and odd-mode (Z_{odd}), respectively. For the analysis of the even-mode, a magnetic wall is positioned on the symmetry plane, as depicted in Fig. 2.34. Conversely, for the analysis of the odd-mode, an electrical wall is placed on the same plane.

The coupling coefficients were initially introduced in Oliver (1954) as a means to assess the velocities at which the even- and odd-modes propagate. In that specific paper, k_L was defined in the following manner:

$$k_L = \frac{L_{even} - L_{odd}}{L_{even} + L_{odd}} \tag{2.26}$$

where L_{even} and L_{odd} represent the overall inductance of the even- and odd-modes, respectively. Similarly, the electric coupling coefficient k_C was defined using the overall capacitance of the even- and odd-modes, C_{even} and C_{odd}, respectively:

$$k_C = \frac{C_{even} - C_{odd}}{C_{even} + C_{odd}} \tag{2.27}$$

Provided that the even- and odd-mode propagation velocities, $v_{\varphi e}$ and $v_{\varphi o}$, can be expressed as:

$$v_{\varphi e} = \frac{1}{\sqrt{L_{even} \cdot C_{even}}} \tag{2.28}$$

and

$$v_{\varphi o} = \frac{1}{\sqrt{L_{odd} \cdot C_{odd}}} \tag{2.29}$$

it can be shown that, if a backward-type coupler is desired (i.e. requiring equal even- and odd-mode phase velocities, $\beta_e = \beta_o$), the two coupling coefficients must be equal and hence $k_L = k_C$.

Finally, the even- and odd-mode characteristic impedances in a lossless coupled S-CPW can be calculated as follows:

$$Z_{even/odd} = \sqrt{\frac{L_{even/odd}}{C_{even/odd}}} \tag{2.30}$$

2.5.4.1 Magnetic Coupling

The behavior of coupled S-CPWs is analyzed in terms of their magnetic coupling. Using the model of Fig. 2.34, even- and odd-mode inductances can be derived as follows:

$$L_{even} = L_S + L_G - 2L_{SG} + L_{SS}, \quad L_{odd} = L_S + L_G - 2L_{SG} - L_{SS} \tag{2.31}$$

which yields a k_L in the following form:

$$k_L = \frac{L_{SS}}{L_S + L_G - 2L_{SG}} \tag{2.32}$$

So, the magnetic behavior of the structure is independent of the shielding structure. This is an expectable behavior since, the addition of ribbons below the strips does not modify the magnetic field.

2.5.4.2 Electric Coupling

Unlike the magnetic coupling, Fig. 2.34 shows that the electric coupling strongly depends on the shielding plane topology. The three cases are presented separately.

Uncut Shield In the case of a coupled S-CPW with an uncut shield, the expression of its even-mode capacitance can be written as:

$$C_{even} = \frac{C_S \cdot C_G}{C_S + C_G} \tag{2.33}$$

On the other hand, the odd-mode capacitance is expressed as follows:

$$C_{odd} = 2C_{SS} + C_S \tag{2.34}$$

C_S is the only capacitor that intervenes in both expressions. Hence, the tuning of C_G and C_{SS} leads to an independent modification of C_{even} and C_{odd}. Note that these capacitances are related to geometrical dimensions and can be tuned independently from each other.

Center-Cut Shield A capacitance is observed in the model between cut ribbons, and noted as $2C_f$, where C_f is the total capacitance between the two parts of the split ribbon – and thus becomes $2C_f$ when the symmetry plane is considered. As for the previous scenario, the even- and odd-mode capacitances can be derived.

Note that for the even-mode capacitance, C_f has the same potential at both of its ends, and hence, it does not affect the even-mode behavior of the coupled lines. In this scenario, the even-mode capacitance of the coupled lines can be written as in (2.33). However, the odd-mode capacitance is different from the previous case. Indeed, C_f does not have the same potential at both of its ends when the odd-mode is excited. Under these considerations, the expression of C_{odd} can be written as follows:

$$C_{odd} = 2C_{SS} + \frac{C_S \cdot (2C_f + C_G)}{C_S + 2C_f + C_G} \tag{2.35}$$

This yields to a modified k_C and a greater odd-mode characteristic impedance, when compared to the uncut case. This extra degree of flexibility is crucial, as this modification has virtually no impact on the magnetic behavior of the structure.

In other words, the uncut structure does effectively allow the modification of C_{even} and C_{odd} by tuning C_G and C_{SS}. This can be done, e.g. with the modification of S, the gap between the two signal strips, or W_G, the ground strip width, (see Fig. 2.31). However, this does modify the magnetic field around the structure. Hence, the addition of the center cut is a key tool for increasing the design flexibility of coupled S-CPWs.

Side-Cut Shield Finally, let us consider a side-cut coupled S-CPW. In this scenario, the capacitive coupling appears between the two sides of the cut ribbon (located on the side of the structure), which can be represented by the capacitor C_f. If a modal analysis is performed, one can notice that for the odd-mode – where a virtual ground is placed at the symmetry axis – C_f is directly connected to ground and hence, does not modify the odd-mode behavior, as compared to the uncut structure. Thus, the expression in (2.34) is still valid. On the other hand, the introduction of a side-cut modifies the even-mode characteristic impedance. The value of C_{even} can be expressed as follows:

$$C_{even} = \frac{C_S \cdot C_G \cdot C_f}{(C_S \cdot C_G) + (C_S \cdot C_f) + (C_G \cdot C_f)} \tag{2.36}$$

In the case of the side-cut shielding, the even-mode capacitance is reduced as compared to the uncut case. For the latter, C_{even} is calculated as the series configuration of C_S and C_G, as shown in (2.33), whereas for the side-cut shield, C_{even} is the equivalent capacitance of the series configuration of C_S, C_G, and C_f. Hence, (2.36) inevitably leads to a lower value than (2.33). As a result, the side-cut shield coupled S-CPWs present a lower even-mode capacitance, a modified k_C and a greater Z_{even}. The odd-mode characteristic impedance is the same as in the uncut case.

To summarize, the previous analysis shows that, for a given lateral dimension of the coupled S-CPW (i.e. given W_S, S, G, and W_G), the different shielding topologies allow to tune k_C, Z_{even} and Z_{odd} independently from k_L. The expressions for the k_L, C_{even}, and C_{odd} are summarized in Table 2.3 for the three considered shielding topologies.

Based on the expressions given in Table 2.3, a comparison on k_L, C_{even}, C_{odd}, k_C, Z_{even}, Z_{odd} can be carried out, as compared to the uncut case. This comparison is presented in Table 2.4, where the symbol $\rightarrow$ stands for an unchanged parameter, $\uparrow$ stands for an increase in the parameter, and $\downarrow$ represents a decrease in the parameter, as compared to the uncut case.

Table 2.3 Summary of the expressions of the magnetic coupling, even- and odd-mode capacitances of coupled S-CPWs.

	Uncut shielding	Center-cut shielding	Side-cut shielding
k_L		$\frac{L_{SS}}{L_S + L_G - 2L_{SG}}$	
C_{even}	$\frac{C_S \cdot C_G}{C_S + C_G}$	$\frac{C_S \cdot C_G}{C_S + C_G}$	$\frac{C_S \cdot C_G \cdot C_f}{(C_S \cdot C_G) + (C_S \cdot C_f) + (C_G \cdot C_f)}$
C_{odd}	$2C_{SS} + C_S$	$2C_{SS} + \frac{C_S(2C_f + C_G)}{C_S + 2C_f + C_G}$	$2C_{SS} + C_S$

Table 2.4 Impact on the main coupler parameters due to the addition of a center-cut or a side-cut shield.

Shield	k_L	C_{even}	C_{odd}	k_C	Z_{even}	Z_{odd}
Center-cut	$\rightarrow$	$\rightarrow$	$\downarrow$	$\downarrow$	$\rightarrow$	$\uparrow$
Side-cut	$\rightarrow$	$\downarrow$	$\rightarrow$	$\uparrow$	$\uparrow$	$\rightarrow$

2.5.4.3 Lossy Model of a Coupled Slow-Wave CPW

The expressions carried out in this section and summarized in Tables 2.3 and 2.4 help to understand the behavior of the structure. Next a lossy model is needed for the proper modeling of coupled S-CPWs.

Figure 2.34 depicts a lossless model that can be expanded to incorporate losses resulting from the inclusion of resistive effects. To simplify matters, and as for the S-CPW, radiation and dielectric losses have been disregarded in this analysis. This simplification holds true for structures with a physical length below λ_G and integrated within a high-Q dielectric, such as coupled S-CPWs in integrated technologies. Therefore, the current proposal extends the model shown in Fig. 2.34 to encompass metallic losses in all the conductors of a coupled S-CPW. The lossy coupled S-CPW model is illustrated in Fig. 2.35. The lossy elements can be calculated in the same manner as for the S-CPW, in Section 2.3.2.2.

2.5.5 Design Charts

By employing the aforementioned model, the designer can build design charts to speed up the design process. The different capacitances, resistances, and inductances can be calculated by means of quasi-static simulations using 3D electromagnetic software or by using the method developed for the S-CPW (Section 2.3.2.2). These calculations, in conjunction with equations (2.26)–(2.36), make it possible to construct charts for k_L, k_C, and Z_C. These charts hold significant value for designers as they facilitate the process of narrowing down the range of geometries that yield the desired values for these parameters.

Since the back-end-of-line of integrated technologies cannot be disclosed, a CMOS-like technology was considered to demonstrate how charts could be implemented and used to design coupled S-CPWs. Let us consider that the technology features an upper metal with a thickness of 3 μm, which is a realistic feature for a technology aimed at mm-waves. In this investigation, the signal and ground strips were constructed using a metal of a 5-μm thickness, which can be achieved in integrated technologies by means of metal stacking. The floating shield was positioned at a distance (denoted as h) of approximately 2 μm with respect to the signal/ground strips. Within this setup, the dimensions S and W_G were kept constant at 2 and 15 μm, respectively. The values of RW and RS were fixed at 0.5 μm. Furthermore, for the side-cut- and center-cut-shielded structures, the parameter CS was set to 2 μm. As an illustrative example, the ratios of G/S and W_S/h vary within the ranges of 10–70 and 1–13, respectively. The operating frequency was established at 100 GHz. The resulting magnitudes of k_L, Z_C, and k_C are plotted in Fig. 2.36. The calculation for the characteristic

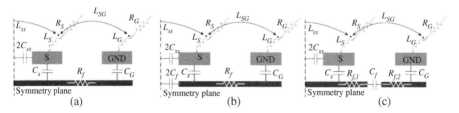

Figure 2.35 Lossy distributed model elementary cell of a lossless coupled S-CPW with central structure symmetry with (a) uncut shield; (b) center-cut shield; and (c) side-cut shield

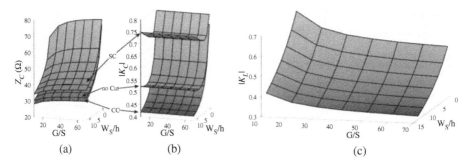

Figure 2.36 (a) Characteristic impedance; (b) electric coupling k_C and (c) magnetic coupling k_L of a coupled S-CPW for the uncut, side-cut (SC) and center-cut (CC) floating shield. ($S = 2\,\mu m$, $W_G = 15\,\mu m$, $RW = RS = 0.5\,\mu m$, and $CS = 2\,\mu m$.)

impedance is as follows:

$$Z_C = \sqrt{Z_{even} \cdot Z_{odd}} \tag{2.37}$$

It is important to note that k_L possesses a singular value for a specific geometry, irrespective of the type of shielding employed, as discussed earlier.

Several notable conclusions can be drawn from the charts presented in Fig. 2.36. However, it is crucial to consider that some of these conclusions rely on the relative magnitudes of the model parameters, specifically the geometries under consideration. While this prevents the formulation of universal rules for coupled S-CPW design, the charts effectively demonstrate the behavior of the structure within a typical back-end-of-line using realistic dimensions.

Firstly, in Fig. 2.36(b), the electric coupling is independent of the variation in G as long as this dimension is sufficiently large to prevent direct coupling between the signal and ground strips, which is the case in our study. Secondly, for the center-cut shield case, the electric coupling decreases with increasing W_S. This can be attributed to the fact that C_{even} increases at a faster rate than C_{odd}, with the latter being dominated by C_{SS}. Conversely, for the uncut and side-cut shield cases, the electric coupling coefficient initially decreases and then increases with W_S/h. This behavior is particularly evident in the side-cut shield topology. For the uncut shield, the decrease can be explained by the fact that for very narrow strips (i.e. small W_S/h ratio), the value of the even-mode capacitance is almost equal to C_S. As C_S starts to increase with W_S/h, C_{even} increases at a faster rate than C_{odd}, resulting in a decrease in k_C. However, as W_S/h continues to increase, the value of C_{even} becomes limited by C_G while C_{odd} continues to increase, leading to an increase in k_C. A similar analysis can be applied to understand the behavior of k_C in the side-cut shield case, where C_G is in series with C_f, causing the change in slope of k_C to occur at smaller values of W_S/h.

The behavior of the magnetic coupling coefficient can be analyzed in a similar manner. As G/S increases, the partial self-inductance of the signal strips (L_S) increases due to the corresponding increase in magnetic flux. Additionally, the partial mutual inductance between the signal and ground strips (L_{GS}) decreases due to the increased distance between them. Conversely, the signal-to-signal partial mutual inductance (L_{SS}) remains unchanged with respect to the G/S ratio. Based on these considerations and equation (2.32), it can be inferred that an increase in G/S results in a decrease in k_L. Furthermore, an increase in W_S/h leads to a decrease in the magnetic coupling coefficient. This can be attributed to the fact that the signal-to-signal partial mutual inductance decreases at a faster rate than L_S, while L_G and L_{GS} remain practically unaffected by the variation in W_S/h. As a result, k_L decreases with increasing W_S/h.

Lastly, the charts in Fig. 2.36 demonstrate that it is possible to independently vary the coupling coefficients k_L and k_C for a given value of the characteristic impedance Z_C. Thus, the inherent versatility of coupled S-CPWs, associated with the additional degrees of freedom provided by variations in shielding topologies, presents an ideal structure for integrated coupler design (see Section 2.6.4.2).

2.6 Circuits Using Slow-Wave CPW and Slow-Wave CPS

At RF frequencies, say below 30 GHz, quite a few lumped components are used in CMOS technologies in order to achieve the circuits needed in RF front-ends. This is because the quality factor of the inductors and transformers is higher or equal to that obtained for transmission lines, with values between 10 and 20, and an evidently higher compactness. When the frequency increases, the surface area between lumped and distributed circuits tends to converge to similar values. For instance, at 60 GHz, only one turn is necessary to achieve the targeted value of planar spiral inductances. Hence, the magnetic flux is only shared by one loop. As a consequence, meandered transmission lines become quite as efficient, in terms of compactness, as spiral inductors. On the other hand, the design of spiral inductors, transformers, and lumped baluns, becomes very tricky, because some parasitic effects (with less impact at lower frequencies) have to be considered with a high accuracy. For instance, the parasitic electrical coupling between primary and secondary windings of transformers leads to a high design complexity that requires the development of very accurate models. This is why this section focuses on the design of high-performance mm-wave passive circuits based on the slow-wave transmission lines. The use of high-performance passive circuits can be a very good lever for achieving high-performance mm-wave active circuits, like low-noise amplifiers (LNAs), power amplifiers (PAs), or voltage-controlled oscillators (VCOs).

When dealing with CMOS/BiCMOS technologies, the strategy for the design of transmission lines differs from the one used when dealing with standard printed circuit board (PCB) technologies. On one hand, the height of the back-end-of-line and design rules impose severe constraints, on the other hand, the multiple metal layers offer great flexibility, in particular when dealing with slow-wave transmission lines. This flexibility can be compared to that of other multi-layer technologies like low temperature co-fired ceramics (LTCC) or 3D Advanced technologies (Tentzeris et al., 2004; Hao & Hong, 2008). In the following, this flexibility is used to achieve several passive circuits.

The passive circuits classically take place in mm-wave front-ends. Since most of them are based on the use of junctions, these are presented first. In particular, the realization of hybrid junctions with S-CPW and microstrip lines is described. Then, different resonator types and various filter topologies are presented. Power dividers and combiners are also addressed. Two types of dividers/combiners are described, with or without isolation between the two output ports. Then couplers based on hybrid and coupled line topologies are presented. Baluns are then discussed through two topologies based on a rat-race and a power divider. Both topologies use CPW phase inverters in order to achieve compact designs with very wide bandwidths in terms of phase difference flatness. A VCO tank is presented with a slow-wave based inductor. Then, different kinds of phase shifters are developed. One is fully integrated, with the frequency control being realized by the use of varactors. Another one benefits from the slow-wave technology to combine liquid crystal tunability with MEMS tunability to get a high phase shift. Finally, a last application is presented, which consists of a sensor for characterization purposes.

2.6.1   Junctions

In order to realize a circuit for a specific application, transmission lines with various characteristic impedances and topologies have to be connected in different ways. As the electrical properties of the concerned transmission lines usually differ, their geometrical dimensions also do. So, the way to connect those transmission lines may influence the circuit behavior and should be considered during the design step. In this section, the straight junction between S-CPWs and microstrip lines is addressed, along with the T-junctions.

2.6.1.1   Microstrip to Slow-Wave CPW Junction

The junction between S-CPWs and microstrip lines is necessary for almost all the circuits utilizing S-CPWs, because the use of both topologies in the same circuit often offers a great flexibility to reach a trade-off between electrical performance and compactness. This kind of junction has a significant effect, mostly on the magnetic field. Actually, if the dimensions are well-chosen, the S-CPW patterned shield acts as a full shielding toward the electric field, which shows a similar distribution as in the microstrip line topology. On the contrary, the magnetic field in the slow-wave propagation mode flows below the shielding, which is not the case for a conventional microstrip line. Figure 2.37 shows a possible schematic for the straight junction. At the junction, the connection between the microstrip ground plane and the S-CPW ground strips is ensured through vias arrays, consisting of a large number of vias to reduce the associated resistance. Vias are mandatory to prevent the odd-mode propagation in the S-CPW.

Figure 2.38 shows the photograph and measurement results of two cascaded 50-Ω straight junctions between μstrip_B9MW and S-CPW6 from Table 2.1. As the return loss is better

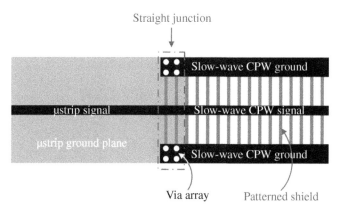

Figure 2.37   Microstrip to S-CPW straight junction schematic.

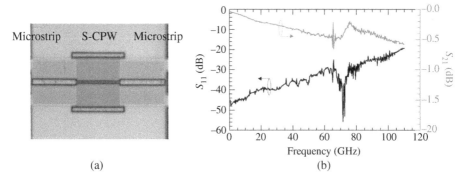

(a)	(b)

Figure 2.38 (a) Photograph and (b) measured S-parameters for a double straight junction microstrip – S-CPW – microstrip.

than 20 dB up to 110 GHz, such a microstrip to S-CPW junction is good enough for future complex design.

2.6.1.2 Tee-Junctions

Particular issues arise for the tee or cross junctions, for any kind of transmission line, even microstrip, for which specific models were developed in the eighties (Hammerstad, 1981; Kirschning & Jansen, 1982). Indeed, the shielding arrangement below the junction is very challenging because two parallel ribbon sets (that might be on the same metal layer) exist in orthogonal directions. Moreover, this phenomenon is enhanced because the S-CPWs usually exhibit large total widths, which means that a slow-wave junction can have a non-negligible electrical length.

Figure 2.39 suggests two types of tee-junctions for S-CPWs. Both require bridges between all the ground planes so that only the even-mode propagates in the structure after the tee, which is necessarily asymmetric. This is easily ensured by dense via groups, which connect

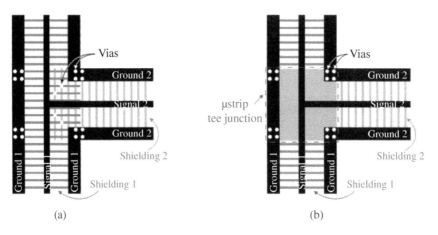

(a)	(b)

Figure 2.39 Two possible tee-junctions for S-CPWs topology using only S-CPWs (a) or the introduction of a microstrip transmission line (b).

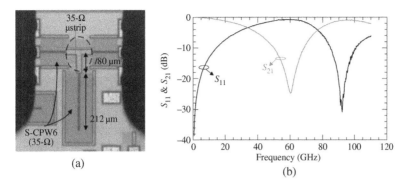

(a)

(b)

Figure 2.40 Slow-wave stub with a microstrip tee junction: (a) Under probe photograph and (b) measured S-parameters.

each ground ribbon with the shielding plane. The solution in Fig. 2.39(a) uses only S-CPWs. Optional vias may also be added at the junction between the orthogonal shielding ribbons if they are implemented on different metal layers. Other variations can be suggested. However, as long as the phase velocity is not precisely known at the junction, this solution leads to a lack of precision. The solution given in Fig. 2.39(b) overcomes this problem by introducing a well-characterized conventional microstrip tee junction. As the connection between S-CPWs and microstrip lines is not an issue (see Fig. 2.38), this solution is the most suited in order to achieve accurate tee junction for S-CPWs.

The solution with a microstrip tee junction was used to implement a 60 GHz resonating stub in the STM B9MW 0.13 μm technology (Fig. 2.40(a)). The tee junction was realized with an 80-μm long 35-Ω microstrip line (signal width of 20 μm). The main part of the stub was realized with a S-CPW also having a 35-Ω characteristic impedance. The whole stub resonates at 60 GHz (Fig. 2.40(b)): The microstrip line introduces an 11° phase shift, whereas the slow-wave part corresponds to 77°, resulting in a quarter-wavelength stub. Note that, if a 35-Ω microstrip line were used for the whole resonator, the total length would have been 2.2 times longer, confirming the slow-wave miniaturization interest.

2.6.2 Millimeter-Wave Filters

2.6.2.1 Dual Behavior Resonator

Based on Franc et al. (2012b), two 60-GHz Dual Behavior Resonator (DBR) filters are detailed in this section, in order to highlight the filter performance achievement with S-CPWs. A classical DBR resonator consists of a transmission line loaded by two parallel open-ended stubs. This topology introduced in Quendo et al. (2003) is shown in Fig. 2.41(a). The whole structure composed of the two stubs behaves like a half-wavelength resonator at the working frequency. Besides, each stub introduces its own transmission zero when its electrical length is equal to a quarter-wavelength. Transmission zeros are located on both sides of the working frequency.

The 60-GHz electrical characteristics of both filters, obtained from the design method described in Quendo et al. (2003), are given in Fig. 2.41.

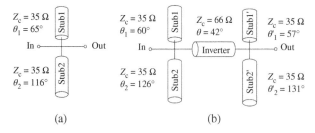

(a)   (b)

Figure 2.41 Topology of (a) a single DBR resonator (first-order DBR filter); and (b) a second-order DBR filter.

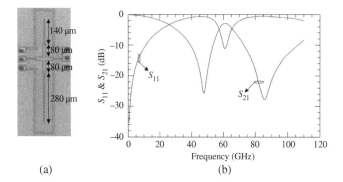

(a)   (b)

Figure 2.42 (a) Photograph of the first-order DBR and (b) measurement results.

A photograph of the first-order DBR filter is shown in Fig. 2.42(a), and the measurement results are given in Fig. 2.42(b). The measured insertion loss (IL) is 2.6 dB. The return loss is better than 10 dB, and the 3-dB fractional bandwidth reaches 18%. These results prove the possibility of designing interesting low-loss resonators based on S-CPWs with a 60-GHz working frequency.

The photography of the second-order DBR filter is shown in Fig. 2.43(a). Measurement results are shown in Fig. 2.43(b). The measured insertion loss reaches 4.1 dB. The 3-dB fractional bandwidth is 17%. The out-of-band rejection between 75 and 105 GHz is higher than 30 dB. Even if some parameters should be improved (like flatness or out-of-band rejection), these measurement results confirm that S-CPWs are likely a good candidate to reach both reasonable insertion loss and a relatively low 3-dB bandwidth.

Note that the DBR structures need cross-junctions, which are made of microstrip based T-junctions.

Finally, a comparison between the use of S-CPWs and microstrip lines is interesting. In this perspective, the simulated results of this second-order filter with the measured characteristics of S-CPWs and microstrip lines are plotted in Fig. 2.44. Simulations are based on models extracted from the 60 GHz measured microstrip lines and S-CPWs. The predicted minimum insertion loss is 2.1 dB when S-CPWs are used, as compared to 4.4 dB for microstrip lines, for almost the same relative bandwidth of 18%. Additionally, even if meandered microstrip lines could be considered, the microstrip filter would still show an area of 0.34 mm^2, whereas the S CPW filter needs only a 0.29 mm^2 area.

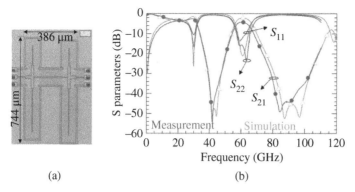

(a)   (b)

Figure 2.43 Second-order 60-GHz DBR filter: (a) photograph; (b) measurement and simulation results.

Figure 2.44 S-parameters simulations of the 2nd order DBR filter designed with either microstrip lines or S-CPWs (based on transmission lines measurement results).

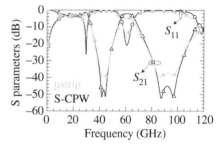

2.6.2.2 Coupled Lines Filters

Another application of the S-CPWs is detailed in the following: A bandpass filter based on coupled S-CPWs is presented and compared to its microstrip counterpart. The bandpass filter topology is derived from the well-known parallel coupled line topology (Hong & Lancaster, 2001), Fig. 2.45. A third-order filter is demonstrated with the implementation of three half-wavelength resonators. From a practical point of view, this arrangement offers the alignment of the input and output access as well as great compactness. On the other hand, it creates inter-resonator cross-couplings through the gaps x_1 and x_2, leading to the generation of transmission zeros. In between those gaps, the filter can be divided into four sections of coupled lines. The calculation of the odd- and even-mode characteristic impedances associated with each section was made using the Chebyshev function with 0.1 dB ripple and a fractional bandwidth of 25%. Table 2.5 summarizes the values of the electrical parameters.

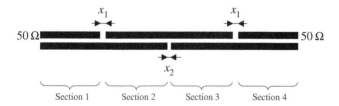

Figure 2.45 Schematic of the third-order bandpass filter.

Table 2.5 Electrical parameters of the bandpass filter with a 25% fractional bandwidth.

ith section	$(Z_{0e})_i$	$(Z_{0o})_i$
1 and 4	103 Ω	38 Ω
2 and 3	78 Ω	38 Ω

The technological implementation in BiCMOS 55 nm technology from STMicroelectronics was carried out for a 120-GHz center frequency in two different topologies for comparison: with coupled S-CPWs and with microstrip coupled lines. The main strips were implemented on the top thick metal layer for both coupled S-CPW and microstrip line designs. The ground plane of the microstrip lines was implemented on the lowest metal layer, and the shielding ribbons of the coupled S-CPWs were placed on an intermediate level to provide both a noticeable miniaturization effect and the correct characteristic impedance range. The photographs of the two filters, shown in Fig. 2.46, highlight the longitudinal miniaturization thanks to the slow-wave effect. Nevertheless, due to a larger width, the coupled S-CPW implementation exhibits a larger area (0.071 mm²) as compared to its microstrip counterpart (0.045 mm²).

Figure 2.47 depicts the measurement and simulation results for the designed bandpass filters. Both filters exhibit a return loss better than 10 dB in the passband. The measured mid-band insertion loss is slightly lower with the coupled S-CPWs compared to microstrip coupled lines (4.9 dB versus 5.2 dB). Moreover, the good flatness achieved by the coupled S-CPWs filter within the passband is the consequence of the design flexibility offered by the coupled S-CPWs structure in the optimization of the even- and odd-mode characteristic impedances.

Thus, an interesting point of the design using coupled S-CPWs is the flexibility offered by a greater number of adjustment parameters: width of the strips W and W_g, gap G, and height between strips and shielding ribbons h for the coupled S-CPW versus only strip width and gap for microstrip coupled lines. Conversely, this means a longer design time for designers.

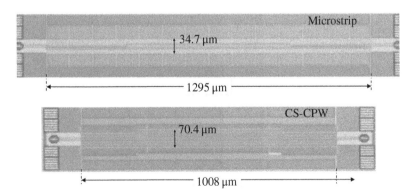

Figure 2.46 Chip photograph of the third-order 120-GHz bandpass filters with microstrip coupled lines and coupled S-CPWs.

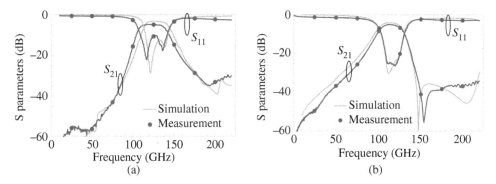

Figure 2.47 Measurement and simulation of the bandpass filters designed with (a) coupled S-CPWs and (b) microstrip coupled lines.

2.6.2.3 LC Quasi-Lumped Resonator

The low-quality factor of the lumped passive components such as inductances and capacitances has always been a concern for radio-frequency integrated circuit (RFIC) design. To mitigate this issue, the S-CPS was advertised as a good candidate to derive high-quality factor quasi-lumped inductances and capacitances. Then, based on those components, very promising LC resonators could be designed with Fig. 2.48, as will be shown in this section.

From the ABCD matrix of the short-circuited transmission line, the input impedance Z_{in_L} can be expressed as:

$$Z_{in\_L} = \frac{B}{D} = Z_c \cdot \tanh(\gamma l) \tag{2.38}$$

with $\gamma = \alpha + j\beta$ the propagation constant, and l the transmission line physical length. Let us first consider the lossless case, i.e. $\gamma = j\beta$, in order to show how the slow-wave concept allows increasing the equivalent inductor L_{eq}. In that case, Z_{in_L} can be written as:

$$Z_{in\_L} = jZ_c \cdot \tan(\beta l) = jL_{eq}\omega \tag{2.39}$$

Thus, the equivalent inductor can be derived as:

$$L_{eq} = \frac{Z_c \cdot \tan(\beta l)}{\omega} \tag{2.40}$$

Note that the shorted transmission line can only be equivalent to an inductor when βl is lower than $\pi/2$ (modulo π). By restricting βl to a value much lower than unity (i.e. a physical

Figure 2.48 Layout and equivalent model for (a) a short-circuited S-CPS; and (b) an open-circuited S-CPS.

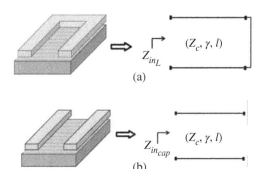

length much lower than the guided wavelength λ_g), a first-order Taylor series development leads to the quasi-static (QS) expression of L_{eq}:

$$L_{eq-QS} = \frac{Z_c \cdot \beta \cdot l}{\omega} = \frac{Z_c \cdot \sqrt{\varepsilon_{reff}}}{c_0} l \tag{2.41}$$

where c_0 is the speed of light in vacuum, and ε_{reff} is the effective relative dielectric constant of the media. If the telegrapher's LC distributed electrical model of the transmission line is considered, the equivalent inductor can be written as (2.42). This means that the equivalent inductor is equal to the linear inductor L of the transmission line multiplied by the physical length when the latter is much lower than the guided wavelength λ_g.

$$L_{eq-QS} = \frac{Z_c \cdot \beta \cdot l}{\omega} = \sqrt{\frac{L}{C}} \cdot \frac{\omega\sqrt{L \cdot C} \cdot l}{\omega} = L \cdot l \tag{2.42}$$

Based on (2.40) and (2.42), the equivalent inductor (L_{eq}) and its quasi-static value (L_{eq-QS}) are plotted versus βl (or similarly versus f). Figure 2.49 shows an example of an equivalent inductor based on a typical classical transmission line and a slow-wave one. For a fair comparison, these two transmission lines have the same characteristic impedance $Z_c = 50\ \Omega$ and physical length $l = 100\ \mu m$. They differ in terms of effective relative dielectric constant ε_{reff}. The effective relative dielectric constant of the classical transmission line is equal to $\varepsilon_{reff-classical} = 4$, whereas it is equal to $\varepsilon_{reff-SW} = 9$ for the slow-wave transmission line. These two values of effective relative dielectric constant correspond to typical cases in (Bi)CMOS technologies.

When βl is much lower than unity, i.e. for low frequency, L_{eq} is equal to L_{eq-QS}. When βl increases (i.e. either the frequency or the physical length increases), L_{eq} increases and differs from the quasi-static case, as shown in Fig. 2.49. When $\beta l = \frac{\pi}{2}$, an infinite value of L_{eq} is obtained due to the tangent function behavior. This corresponds to a frequency of 362 GHz for the classical transmission line and 250 GHz for its slow-wave counterpart.

L_{eq-QS} is interesting to be considered for comparison purposes. At lower frequencies, for the classical transmission line and its slow-wave counterpart, $L_{eq-QS} = 33.3$ pH and $L_{eq-QS}^{SW} = 50$ pH, respectively. The ratio $\eta_{QS} = \frac{50}{33.3} = 1.5$ corresponds to the slow-wave factor $SWF = \frac{\sqrt{\varepsilon_{reff\text{-}slow-wave}}}{\sqrt{\varepsilon_{reff\text{-}conventional}}}$. Hence, slow-wave transmission lines allow the physical length

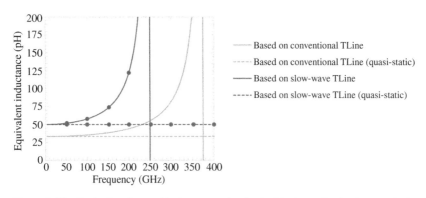

Figure 2.49 Equivalent lumped inductor synthesized with short-circuited transmission line.

reduction to achieve the same equivalent inductor as compared to its classical transmission line counterpart. This phenomenon is magnified by the increase of L_{eq} with frequency. Indeed, as shown in Fig. 2.49, for a working frequency equal to 80 GHz, $L_{eq} = 34.6$ pH for the classical transmission line, and $L_{eq-SW} = 54.7$ pH for its slow-wave counterpart. This leads to a ratio $\eta = 1.58 > \eta_{QS}$.

Let's now consider a lossy transmission line in order to derive the quality factor of the equivalent inductor L_{eq}. As already stated, in that case, a lumped model is used for the shorted transmission line composed of L_{eq} in series with R_{eq}. The equivalent quality factor Q_{eq} is given by:

$$Q_{eq} = \frac{L_{eq} \cdot \omega}{R_{eq}} \tag{2.43}$$

and $tanh(\gamma l)$ can be expressed as:

$$\tanh(\gamma l) = \frac{\tanh(\alpha l) \cdot [1 + (\tan(\beta l))^2]}{1 + [\tanh(\alpha l) \cdot \tan(\beta l)]^2} + j\frac{\tan(\beta l) \cdot [1 - (\tanh(\alpha l))^2]}{1 + [\tanh(\alpha l) \cdot \tan(\beta l)]^2} \tag{2.44}$$

Hence, Q_{eq} can finally be written as:

$$Q_{eq} = \frac{\tan(\beta l) \cdot [1 - (\tanh(\alpha l))^2]}{\tanh(\alpha l) \cdot [1 + (\tan(\beta l))^2]} = \frac{\tan(\beta l)}{\tanh(\alpha l)} \cdot \frac{1 - (\tanh(\alpha l))^2}{1 + (\tan(\beta l))^2} \tag{2.45}$$

The same values for Z_c, physical length l and effective relative dielectric constants $\varepsilon_{reff\_conventional}$ and $\varepsilon_{reff\_slow-wave}$, respectively, as for the calculus of L_{eq} in the lossless case, were considered. The attenuation constant is modeled by $\alpha = \alpha_{DC} + \alpha_f \sqrt{f/f_0}$ with typical values $\alpha_{DC} = 0.2$ dB/mm, $\alpha_f = 0.4$ dB/mm, and $f_0 = 60$ GHz. This model corresponds to an attenuation constant equal to 0.6 dB/mm at 60 GHz, which is close to state-of-the-art transmission lines in (Bi)CMOS back-end-of-lines. The equivalent quality factor Q_{eq} is plotted in Fig. 2.50 for both cases, i.e. classical (a) and slow-wave (b) transmission lines.

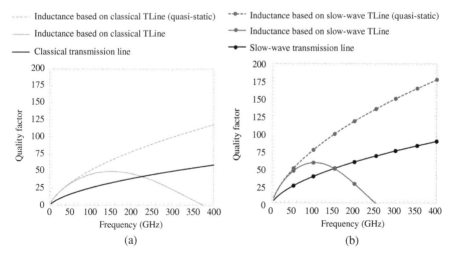

Figure 2.50 Q-factor versus frequency for $Z_c = 50\,\Omega$, $l = 100\,\mu m$: (a) for classical transmission lines ($\varepsilon_{reff\_conventional} = 4$) and (b) for slow-wave transmission lines ($\varepsilon_{reff_{slow}-wave} = 9$).

When considering low-loss transmission lines, the quasi-static quality factor (at low frequency) of the equivalent inductor gives the limit of the maximum quality factor that can be achieved. From (2.45) and for $\alpha l \ll 1$ and $\beta l \ll 1$, $Q_{eq-QS} = \frac{\beta}{\alpha}$ is easily derived. It is also plotted in Fig. 2.50 for comparison. Note that it is twice the well-known quality factor of a transmission line $Q_{TL} = \frac{1}{2}\frac{\beta}{\alpha}$.

Q_{eq} follows Q_{eq-QS} up to a frequency that is roughly equal to 20% of the resonance frequency of the shorted transmission line (e.g. when its electrical length is equal to 90°). The maximum Q_{eq} is obtained at a lower frequency for the slow-wave transmission line, i.e. 104 GHz, as compared to 152 GHz for its classical transmission line counterpart, respectively. Moreover, the maximum Q_{eq} is greater for the slow-wave transmission line, i.e. 57.7 as compared to 49.6, respectively.

In a similar way, the capacitance of a loss-less open-circuited transmission line can be calculated, resulting in (2.46). Taking the losses into account, the quality factor of the equivalent capacitance is expressed as the quality factor of the inductance $Q_{eq-QS} = \frac{\beta}{\alpha}$.

$$C_{eq} \approx \frac{l \cdot \sqrt{\varepsilon_{reff}}}{c_0 \cdot Z_c} = C_s \cdot l \tag{2.46}$$

These simple calculi show that slow-wave transmission lines lead to, (i) shorter stubs for the realization of equivalent inductors and capacitors, and (ii) higher quality factors for those lumped elements, as compared to classical transmission lines. The designer can then tune the various geometrical parameters of the S-CPS in order to reach a given value of capacitance or inductance while trading-off to keep high-quality factor.

Using these S-CPS based components, various short-circuited LC parallel resonators were designed and implemented in BiCMOS-55 nm process. Figure 2.51 shows the chip photograph of the resonator and its associated electrical model with the parallel connection of the inductance and capacitance. The resonator is fed using a 50 Ω microstrip T-junction. Then, the short-circuit is created by connecting S-CPS to the ground plane through a stack of vias with high density.

Figure 2.52 illustrates the measured and the simulated responses of the S-CPS resonator. The insertion loss is equal to 0.7 dB, and the return loss is better than 25 dB at the central frequency.

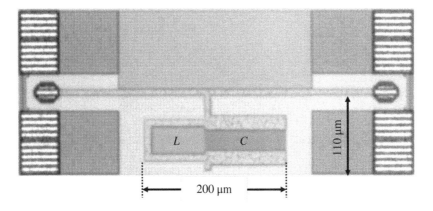

Figure 2.51 Chip photo of the short-circuited parallel LC resonator.

Figure 2.52 Response of the short-circuited parallel LC resonator.

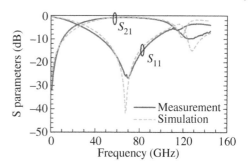

2.6.3 Power Divider/Combiner

2.6.3.1 Wilkinson Topology

A Wilkinson power divider is described in this section. It was implemented in the BiCMOS9-MW technology from STMicroelectronics. A 60-GHz working frequency was targeted with a very conventional design based on S-CPWs having a characteristic impedance Z_0 connected with two parallel quarter-wavelength S-CPWs with a $\sqrt{2} \cdot Z_0$ characteristic impedance. As no isolation resistor was implemented, this power divider is not suited for combining purposes. In the example that follows, Z_0 is equal to $50\,\Omega$. Then, the design flow includes the choice of a T-junction type in order to connect the three branches of the power divider. An optimization of the complete power divider was performed to take the disturbance of the T-junction into account. The simplest solution consists of optimizing the complete device with a 3D electromagnetic tool. However, due to the complexity of the S-CPWs, it requires large computing resources. Another way consists in simulating the T-junction alone (whether slow-wave or microstrip) with a 3D electromagnetic tool, then using the resulting S-parameters in a circuit simulator in order to optimize the electrical length of the two quarter-wavelength transmission lines.

The resulting layout is presented in Fig. 2.53. In this case, a S-CPW T-junction was used for the power divider in BiCMOS9-MW technology, but the design is usually easier with a microstrip type T-junction (Section 2.6.1.2). Two ground–signal–ground probes were used, and port 3 was connected to a $50\,\Omega$ resistance realized in the front-end-of-line, which does not allow to fully characterize the three-port network. The occupied surface of the power divider in BiCMOS9-MW technology is $0.11\,\text{mm}^2$.

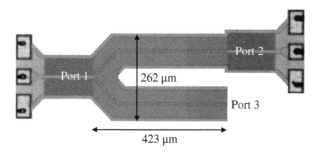

Figure 2.53 Layout of the power divider with S-CPWs in BiCMOS9-MW technology.

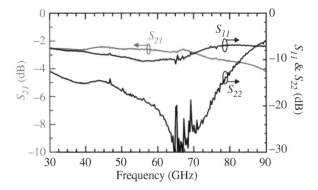

Figure 2.54 Measurement results of the Wilkinson power divider without isolation branch.

For the measurement, a LRRM calibration was first carried out by using an external calibration kit, and then a classical de-embedding method described in Koolen et al. (1991) was used to remove the connecting effect (pad and feeding lines). The matching is about 10 dB from 53 to 64 GHz, Fig. 2.54. The transmission coefficient is as small as −2.7 dB at 60 GHz. In a perfect power divider, a transmission coefficient higher than −3 dB cannot be reached. The major defect comes from the real value of the resistance loading port 3, which was realized in the polysilicon layer and can be estimated between 40 and 60 Ω, due to process dispersion. A value of 40 Ω is sufficient to explain an unbalanced power splitting of 1 dB between the two output ports 2 and 3, leading to more than half of the power flowing through port 2, and less through port 3.

2.6.3.2 Variation Based on Wilkinson Topology

When a power divider is used as a combiner, good output ports matching and isolation are needed, and an isolation resistance is then required, as proposed by Wilkinson (1960). This classical topology may be modified to overcome some classical limitations. Indeed, the undesired coupling between the two quarter-wavelength branches, or the distributed effects introduced by the physical length requirements when stretching the resistance between the two branches can be devastating for the performance of the power divider at mm-waves. Another disadvantage of the classical Wilkinson power divider is the $\sqrt{2} \cdot Z_0$ characteristic impedance of the quarter-wavelength branches. This condition leads to a characteristic impedance of 70.7 Ω for a classical 50 Ω system, which may not be reachable in all CMOS back-end-of-line for microstrip lines. In the case of S-CPWs, however, a 70.7 Ω characteristic impedance is easily reachable, but it exhibits a lower quality factor as compared to the S-CPW with a characteristic impedance around 45 Ω. Hence, flexibility in the choice of the characteristic impedance could permit choosing the transmission lines with the highest quality factor. Moreover, if different characteristic impedances leading simultaneously to full matching, good isolation, and low loss could be found with electrical lengths shorter than the conventional 90°, solutions would result in a more compact device.

Figure 2.55 shows a power divider topology that allows for addressing the issues discussed previously. Compared to the conventional Wilkinson power divider, an open stub of characteristic impedance Z_2 and electrical length θ_2 was added at the junction between the input port 1 and the branches joining output ports 2 and 3. The characteristic impedance and

Figure 2.55 Modified power divider with open stub and resistance feeding lines.

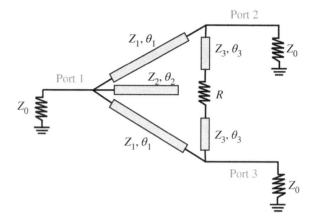

electrical length of the latter were named Z_1 and θ_1. Also, feed lines named Z_3 and θ_3 were added to connect the output ports to the isolation. This combination of Horst et al. (2007) and Ahn (2011) offers a very high flexibility in terms of possible characteristic impedance values and electrical lengths.

In Burdin et al. (2016), an even–odd modes analysis considering the circuit symmetry was applied, leading to six equations. For given port conditions and a given resistor R, the topology totalizes six unknown parameters corresponding to the characteristics of the three transmission lines (Z_i, θ_i), and it can thus be theoretically solved. Nevertheless, some equations are too complex to be solved in an algebraic way, so Burdin et al. (2016) detail a design procedure based on an algorithm. The designer eventually chooses the best solution and can thus take θ_i and/or Z_i possible ranges into account. The designer can thus choose a compromise between the power divider electrical performance and its size by electing the appropriate values of the characteristic impedances and electrical lengths.

The design flow was used to implement a power divider at 60 GHz in BiCMOS 55 nm technology from STMicroelectronics with a resistance R fixed to 105 Ω. The solution $Z_1 = 83\,\Omega$, $\theta_1 = 59°$, $Z_2 = 26\,\Omega$, $\theta_2 = 24°$, $Z_3 = 26\,\Omega$, and $\theta_3 = 15°$ was selected because it leads to a compact device. S-CPWs were used to implement the transmission lines (Z_1, θ_1) between the input and output ports, whereas microstrip lines were chosen for the open stub and the resistance feeding lines because they are narrower and could be easily embedded inside the device. This choice between S-CPWs and microstrip lines corresponds to a trade-off between electrical performance and compactness. The bends and junctions were realized with microstrip lines to make the design easier.

Figure 2.56(a) shows the schematic of the power divider with its physical dimensions. Note that the 83 Ω characteristic impedance cannot be achieved with a microstrip line in the considered technology, so a value of 72 Ω was used, which required a slight optimization to shift back the working frequency to 60 GHz, and to improve the input port matching. The layout is given in Fig. 2.56(b). Differential ground signal ground signal ground (GSGSG) pads were used in order to measure all the S-parameters. The surface area of the circuit is 0.104 mm², and according to the layout, it could be reduced with narrower S-CPWs. With a width of 100 μm instead of the 124 μm used, the total surface could be further lowered to 0.09 mm².

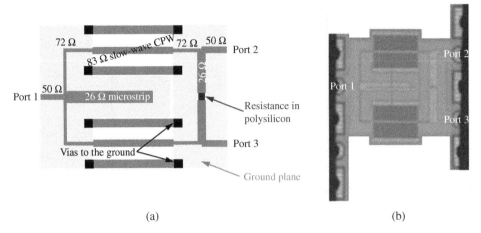

(a)   (b)

Figure 2.56 Modified power divider: (a) schematic with dimensions (not to scale); and (b) layout in the 55 nm BiCMOS BEOL.

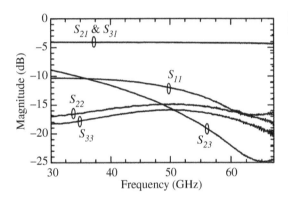

Figure 2.57 Stub-loaded power divider measurement results.

The measurement results in Fig. 2.57 show that the input port return loss is better than 10 dB in the whole measured band, i.e. between 30 and 67 GHz. The isolation is better than 20 dB, and the output port return loss is better than 15 dB between 57 and 67 GHz. The added insertion loss is between 1.2 and 1.5 dB in this bandwidth. Small deviations appear compared to expected results: The frequency shift might be due to the fact that the microstrip to S-CPW junctions were considered perfect in the design step, and the shift of the output port return loss and isolation is justified by the process dispersion, which leads to a polysilicon resistance ranging from 80 to 100 Ω.

The measurement result and the surface area of the modified power divider confirm that the proposed topology is well suited for CMOS mm-wave power dividers. It ensures a low-loss, full-matched, and isolated component together with a small surface, thanks to the use of S-CPWs with high-quality factors. More than that, it emphasizes the importance of flexible topologies, including flexible basic cells like transmission lines, to simultaneously reach compact and high performance circuits.

2.6.4 Couplers & Baluns

2.6.4.1 Branch-Line Couplers

In this section, a conventional branch-line coupler with $-3\,dB$ at each output port is implemented at 60 GHz with S-CPWs in the CMOS bulk 28-nm technology. The system impedance being chosen as $Z_0 = 50\,\Omega$, S-CPWs with characteristic impedances Z_0 and $\frac{Z_0}{\sqrt{2}} = 35\,\Omega$ are required. This is a good point, as the S-CPWs with those characteristic impedance values usually exhibit a high-quality factor and an interesting slow-wave factor. It is then possible to envision compact and low-loss branch-line couplers. Once again, the T-junctions have been designed in microstrip to simplify the simulations and layout design.

Figure 2.58 shows the layout of the circuit, which is fed thanks to GSGSG probes connected to conventional microstrip lines. After the de-embedding step, the ports are defined as shown in Fig. 2.58.

The simulation results are presented in Fig. 2.59. The device has two symmetry axes (vertical and horizontal) so the expected return loss at the working frequency is around 30 dB for the four ports. The insertion loss is low, as the simulated S_{21} and S_{31} are respectively -3.67 and $-3.48\,dB$. The maximal magnitude imbalance is 0.19 dB in the bandwidth defined with a 12 dB return loss. The variation of S_{31} is less flat than the variation of S_{21} because the electrical length between ports 1 and 3 is 180° while it is only 90° between ports 1 and 2, resulting in a stronger narrow-band effect of the quarter-wave length transmission line for S_{31}. The phase difference ranges between 89.6° and 91° in the bandwidth, leading to a phase imbalance of $\pm 0.7°$.

2.6.4.2 Coupled Line Couplers

In this section, four implementations of quarter-wavelength couplers are presented. All these devices were designed using 55-nm BiCMOS technology from STMicro-

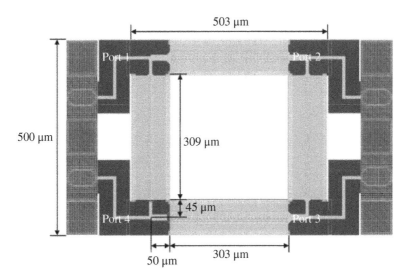

Figure 2.58 Layout of the branch-line coupler with S-CPWs in CMOS bulk 28-nm technology.

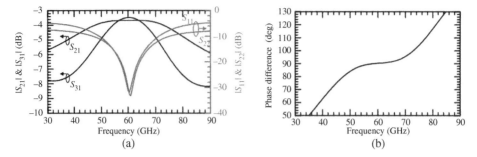

Figure 2.59 Simulation results of the branch-line coupler with S-CPWs implemented in 28 nm technology: (a) Magnitude of the S-parameters and (b) phase difference.

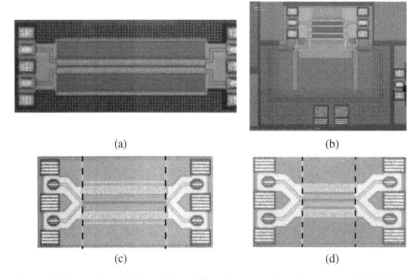

Figure 2.60 Implementation of four different couplers based on coupled S-CPW in a BiCMOS 55-nm technology: (a) 3-dB coupler at 60 GHz; (b) 18-dB coupler at 150 GHz in the configuration of S_{11} and S_{21} measurement; (c) 3-dB coupler at 120 GHz; and (d) 3-dB coupler at 185 GHz.

electronics. The structure used for their implementation is the coupled S-CPWs presented in Section 2.5.

Figure 2.60 presents the microphotographs of the four fabricated couplers based on coupled S-CPW. The first coupler (a) is designed to have a central frequency of 60 GHz. On the other hand, coupler (b) is designed to present a weak 18-dB coupling at 150 GHz. Finally, couplers (c) and (d) are both designed with a 3-dB coupling at 120 and 185 GHz, respectively. Note that coupler (b) was integrated with two tunable loads to allow a wideband measurement (i.e. synthetization of a wideband 50-Ω load). These four couplers were designed with a 50-Ω characteristic impedance.

Table 2.6 presents the electrical performance of the couplers shown in Fig. 2.60. First, note that the 3-dB couplers present large return loss and low magnitude imbalance (<1 dB) between the through and coupled outputs. The couplers at 120 GHz and 185 GHz, whose

Table 2.6 Electrical performance of four couplers based on coupled S-CPW.

Coupler	RL (dB)	Through (dB)	Coupling (dB)	Isolation (dB)	Phase imbalance (°)	Footprint $(\lambda^2)^{a)}$
3-dB 60-GHz[b)]	>20	3.4	3.4	>20	N/A	0.014
18-dB 150-GHz[c)]	>8	2	18	>20	N/A	0.02
3-dB 120-GHz[c)]	>25	3.7	3.6	>25	1	0.003
3-dB 185-GHz[c)]	>15	2.9	3.7	>15	1	0.0045

a) Calculated as the ratio between the footprint and the squared wavelength of an electromagnetic wave traveling in the free-space at the central frequency of the couplers.
b) Simulated.
c) Measured.

phase imbalance was characterized, also show reduced phase imbalance (i.e. good quadrature) between the coupled and through outputs.

The case of the 150-GHz coupler is particular. Indeed, this coupler had two variable loads integrated into two of its outputs. This approach was considered in order to have a wideband 50-Ω load, which is quite a difficult task at such a high frequency. Even though the introduction of this device allowed for the desired coupling level and good isolation, the insertion loss (at the through port) and the return loss were quite degraded.

Finally, the four couplers based on coupled S-CPW present a reduced footprint. This is especially true for the two last couplers that were designed using a small gap G dimension.

2.6.4.3 Rat-Race Balun

A classical rat-race balun is based on transmission lines whose lengths are $\lambda_g/4$ or $3\lambda_g/4$ (λ_g being the guided wavelength), leading to cumbersome circuits, even when slow-wave transmission lines are considered. So, the miniaturization of rat races is of high interest. In this context, Mandal and Sanyal (2007) report that an infinite number of electrical lengths exist for the design of a 3-dB hybrid coupler. Whatever the branches characteristic impedance is, below the conventional value of $Z_0\sqrt{2}$, where Z_0 is the system impedance, the ring electrical length may be less than $3\lambda_g/4$. So, it becomes possible to shorten the rat-race ring, and the characteristic impedance of the S-CPWs can be chosen in order to maximize the quality factor. Moreover, in order to reduce even more the ring physical length (especially to equate the insertion loss) and to improve the phase imbalance between the two output ports over a wide bandwidth, a phase inverter (Chirala & Floyd, 2006) can be inserted in the longest branch of the rat-race.

One implementation in a CMOS 65-nm technology from STMicroelectronics is shown in Fig. 2.61. A characteristic impedance of $Z_c = 45\ \Omega$ was selected for the rat race because it leads to the highest reachable quality factor in the technology. According to Mandal and Sanyal (2007), considering a 45-Ω characteristic impedance, the electrical lengths of the rat-race paths become 52° and 232° instead of 90° and 270° for the conventional topology, respectively. Thus, while choosing the transmission line with the highest quality factor, not only does the rat-race present better loss performance, but also it is more compact.

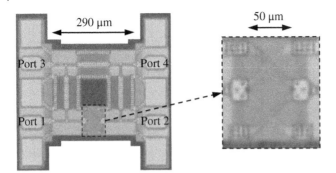

Figure 2.61 Photograph of the rat-race including a phase inverter.

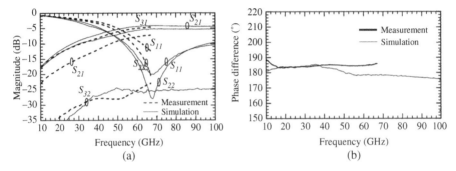

Figure 2.62 Measurement and simulation results of the rat-race including a phase inverter: (a) S-parameters magnitude and (b) phase difference.

Figure 2.62 compares measurement and simulation results. Measurements were carried out from 10 to 67 GHz. The working frequency fits between simulation and measurement results. At 67 GHz, S_{11} and S_{22} reach −12.5 and −20.5 dB, respectively; see Fig. 2.62(a). The transmission coefficient S_{31}, through the branch without phase inverter, fits well with the simulated one and reaches -4.6 dB, which means 1.6 dB of added insertion loss. The transmission through the phase inverter, S_{21}, is equal to −7.2 dB, that is, 4.2 dB of added insertion loss. The transmission parameters are robust, with very flat curves. The S_{11} and S_{21} parameters, which are related to a branch connected to the phase inverter, are degraded as compared to the S_{22} and S_{31} parameters, emphasizing the effects of the phase inverter. This highlights that the phase inverter should be carefully designed to reach high performance in terms of matching and insertion loss. Fig. 2.62(b) shows a phase difference centered around 185° with a phase imbalance of ±1° between 32 and 67 GHz, or 185.5 ± 0.5° in the frequency band from 62 to 67 GHz. This result is obtained thanks to the use of the phase inverter because its phase variation versus frequency is negligible. This leads to an equal phase shift of both rat-race branches over the frequency, because each branch has the same physical length. There is a shift of 5° compared to the perfect phase imbalance, but it is very flat over the whole bandwidth. 5° would be easy to overcome thanks to transmission line length readjustment. Finally, the chip is compact, with an area of 0.085 mm². This limited footprint is due to the inverter, the use of non-50-Ω transmission lines, and a slow-wave effect that allows decreasing the length by a factor of 1.9 as compared to 45 Ω microstrip lines.

2.6.4.4 Power-Divider-Based Balun

The main disadvantage of the rat-race when integrated into a CMOS process is that the orientation of the device is limited to only two positions in quadrature. Therefore, the output ports are far from each other and need non-symmetric interconnects toward the next circuit. Non-symmetric interconnects mean different insertion loss, which can be an issue for many applications. To address this issue, another topology of balun can be used: the classical quarter-wavelength power divider with a phase inverter inserted in one of the output branches to get 180° phase difference between the two output ports that are in the same physical plane. The isolation resistance R used in the Wilkinson power dividers must be removed; otherwise, the 180° relative phase difference would significantly increase the losses. By removing this resistance, this topology cannot be used anymore as a combiner because isolation and output port matching are degraded. A balun based on this concept was designed and fabricated using 65-nm CMOS technology from STMicroelectronics at a working frequency of 60 GHz.

The photograph of the balun is given in Fig. 2.63. The area is equal to 0.1 mm². The output port position is adjusted to fit with the RF probes. In a system including this power divider, the output transmission lines could be designed closer to reduce the surface area.

Measurements from 10 to 67 GHz are presented in Fig. 2.64. The realized bandwidth satisfying a 20-dB return loss is 25% (between 46.5 and 60 GHz). In this frequency band, the measured phase difference is 173.5°, which means a discrepancy of 6.5° compared to the targeted 180°. The phase imbalance is only ±0.4° between 46.5 and 60 GHz. The difference in magnitude between S_{21} (−4.5 dB) and S_{31} (−3.6 dB) is due to the impedance mismatch seen from the input port because the phase inverter acts as a stepped impedance, creating a discontinuity for the signal between ports 1 and 2. This could be improved with a better design.

2.6.5 Voltage-Controlled Oscillator tank

Voltage-controlled oscillators are essential devices for telecommunication circuits and radars because they make it possible to set the working frequency very precisely.

Figure 2.63 Photograph of the power divider balun based on S-CPWs.

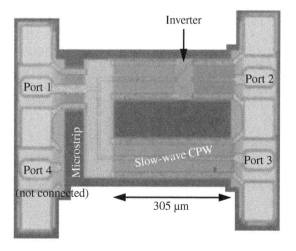

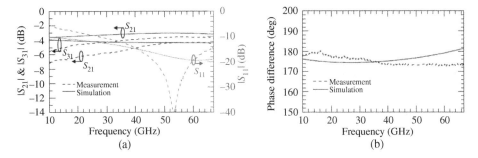

Figure 2.64 Simulation and measurement results of the power divider balun in the 65 nm technology: (a) S-parameters magnitude and (b) phase difference.

The performance of a VCO is mainly defined by the quality of its resonator or tank. At radiofrequency, resonators are essentially made using lumped components, inductors (or transformers), and varactors. As the frequency increases, the design of the inductors becomes more and more complex. Their intrinsic quality factor increases with frequency, as for transmission lines, but their parasitic effects become more and more difficult to consider (in particular the capacitances between turns and with respect to bulk), requiring extremely complex and accurate models. Thus, between 50 and 100 GHz, competition arises between inductors and transmission lines. Beyond about 100 GHz, the advantage is clearly in favor of the transmission lines. Their length becomes more and more acceptable, and they hardly suffer from any parasitic elements.

In silicon technologies, only the limitations imposed by the density of metals for each layer of the BEOL can possibly make them a bit complex to optimize. On the other hand, they fully benefit from the interest of thick BEOLs, which are available for BiCMOS technologies with the aim of improving the performance of passive circuits. The most limiting factor of the VCO performance is the varactor, whose quality factor decreases drastically with frequency. While it is very good at low frequencies, greater than 100 at 1 GHz, it drops to less than 8 around 100 GHz and less than 5 beyond 150 GHz, whatever the CMOS/BiCMOS technology or the technological node considered.

This section shows how S-CPS can be used for the realization of resonators entering into the composition of VCOs at mm-waves.

2.6.5.1 Slow-Wave CPS as Inductor Voltage-Controlled Oscillator

The work presented in this section is detailed in Sharma et al. (2020). The basic principle is illustrated in Fig. 2.65, where the scheme of the VCO is given. The VCO is composed of three parts, namely the tank, the negative resistance (made here by a crossed-coupled pair (CCP)), and the output buffers. The classic tank inductor, generally made in a spiral, has been replaced by a S-CPS shorted at its end (details in Section 2.6.2.3 entitled LC quasi-lumped resonator).

In order to show the consequence of the use of a S-CPS inductor on the performance of a VCO, two VCOs working around 80 GHz were designed and characterized: a reference VCO using a classic spiral inductor of the technology public design kit (PDK), and a VCO using a S-CPS, as illustrated in Fig. 2.65. The design details of the shorted S-CPS are given in Sharma et al. (2020). The BiCMOS 55-nm technology from STMicroelectronics was used for

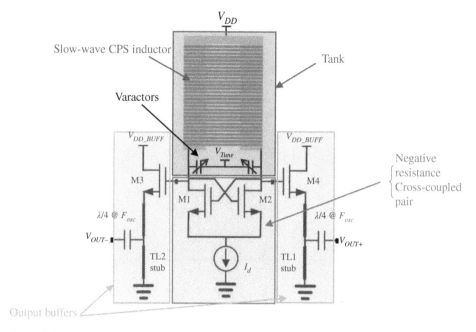

Figure 2.65 VCO schematic with a short-circuited transmission line as inductor.

the fabrication. To achieve a fair comparison, both designs considered the same varactors and the same inductor value of 107 pH. The used lumped inductor is a one-turn spiral shape coil with a width of 5.5 μm and an inner diameter of 55 μm. The cross-coupled pair was sized to compensate the losses of the tanks with the same compensation factor ($K = 3$). Since the losses of the synthesized equivalent inductor are lower (tank parallel resistance $R_p = 966\,\Omega$) than its lumped counterpart ($R_p = 630\,\Omega$), the cross-coupled pair was designed with smaller MOS transistors than the ones of the reference VCO, which leads to lower parasitic capacitances. In addition, it consumes less DC current as presented in Table 2.7.

Figure 2.66 shows the chip micrograph of both VCOs. The area of the synthesized equivalent inductor is comparable to the size of the inductor of the PDK, hence higher electrical performance was achieved without sacrificing the size.

Figure 2.67 shows the variation of the measured oscillation frequency versus the voltage control for both VCOs characterized by on-wafer probing.

The two VCOs do not have quite the same operating frequency, with an offset of around 1.5 GHz. The frequency tuning range (FTR) of the S-CPS VCO is slightly larger than that of the VCO using a conventional inductor, i.e. 3.69 GHz compared to 3.34 GHz. The most noticeable improvement in performance is illustrated by the phase noise, shown in Fig. 2.68.

Another important performance parameter is the power consumption. With a V_{DD} of 1.2 V, the power consumptions of the reference VCO and S-CPS inductor based VCO cores are 9.0 and 5.7 mW, respectively, while the output buffer consumption is 9.2 mW. After correcting the probe and cable losses, measured output powers of reference VCO and S-CPS inductor based VCO are −4.4 and −4.5 dBm, respectively, leading to power efficiencies of 4.0 and 6.2%, respectively.

Table 2.7 Design parameters of the VCOs.

Parameters	LC-tank VCO	VCO with a S-CPS inductor
Inductor	107 pH	107 pH
Varactor capacitance (max/min)	8.9 fF/4.4 fF	8.9 fF/4.4 fF
Varactor max. Q-factor	13.5	13.5
Tank Q-factor	8.5	11.1
Negative Resistance ($R_p/3$)	210 Ω	322 Ω
Cross-Coupled Pair (W/L)	17 μm/0.06 μm	11 μm/0.06 μm
Parasitic Capacitance ($C_{par-CCP}$)	23.5 fF	16.5 fF
Measured bias current (I_d)	7.57 mA	4.74 mA
V_{DD}	1.2 V	1.2 V

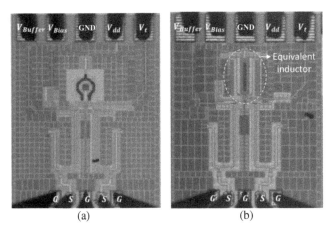

(a)   (b)

Figure 2.66 Chip micrograph of VCO with (a) LC-tank and (b) S-CPS synthesized inductor.

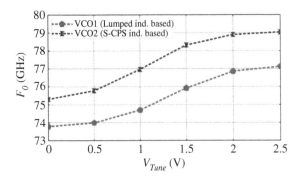

Figure 2.67 Measured oscillation frequency versus V_{Tune}.

Figure 2.68 Measured phase noise of the reference VCO and S-CPS inductance based VCO.

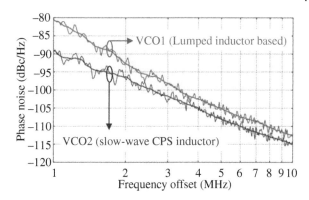

Table 2.8 Measured performance of the VCOs.

VCO type	F_{osc} (GHz)	FTR (%)	P_{DC} (mW)	Phase noise @ 10 MHz (dBc/Hz)	FOM (dBc/Hz)	FOM_T (dBc/Hz)
Reference VCO	75.45	4.43	9.08	−112	−180.5	−173.4
S-CPS inductor based VCO	77.1	4.79	5.69	−115	−185.0	−178.6

To get an overall comparison of the performance of the two VCOs, the Figure of Merit (FOM) and FOM with tuning range (FOM_T) defined as (Lu et al., 2013) were used:

$$FOM = L(\Delta f) - 20 \cdot log \left(\frac{F_{osc}}{\Delta f} \right) + 10 \cdot log \left(\frac{P_{diss}}{1mW} \right) \qquad (2.47)$$

$$FOM_T = FOM - 20 \cdot log \left(\frac{FTR}{10} \right) \qquad (2.48)$$

where $L(\Delta f)$ is the phase noise at the offset frequency Δf from oscillation frequency F_{osc} in dBc/Hz, and P_{diss} is the DC power consumption.

The comparison of the two VCOs is given in Table 2.8. The use of the S-CPS inductor leads to higher FOM and FOM_T.

2.6.5.2 Slow-wave CPS resonator standing wave Voltage-Controlled Oscillator

The use of standing wave resonators instead of lumped LC tanks was initially proposed to increase the resonator Q (Andress & Ham, 2005). However, here again the overall quality factor is still limited by the varactors, and the performance of VCOs based on standing wave resonators is not better than the ones using classical LC Tank. On the other hand, if the design is correctly carried out by integrating the capacitances of the varactors as well as the parasitic capacitance of the cross-coupled pair within the transmission line used as a resonator, the FTR can be greatly improved, which gives standing wave resonators VCOs an undeniable interest.

In this section, S-CPS are used to realize a standing wave VCO. Details of the theoretical approach can be found in (Gomes et al., 2021). The principle of CPS resonator standing wave VCO is illustrated in Fig. 2.69.

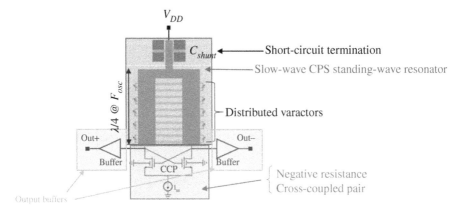

Figure 2.69 S-CPS resonator standing wave VCO.

The output buffers as well as the cross-coupled pair are realized in a similar way to the LC-VCO in Fig. 2.65. On the other hand, the tank is made by a quarter-wavelength S-CPS terminated in a short-circuit, thus forming a resonator. The S-CPS is tunable, with varactors being periodically distributed throughout its entire length. A 3D view of the S-CPS periodically loaded by the varactors as well as the electrical model of the assembly are given in Fig. 2.70.

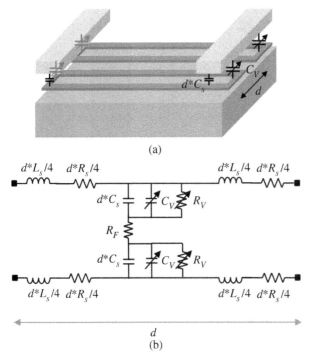

Figure 2.70 (a) 3D view of the varactor-loaded S-CPS; (b) lumped electrical model containing the varactor equivalent circuit for an elementary section of length *d*

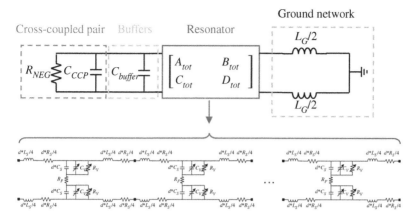

Figure 2.71 Standing wave VCO model with all major components.

As explained in (Franc et al., 2013b) and in Section 2.4.2, the S-CPS is modeled by an $R_sL_sR_fC_s$ circuit. Here, L_s and R_s are the series elements, and C_s and R_f are the shunt elements. The varactors are simply modeled by a variable capacitance C_v in parallel with a resistance R_v.

Based on the S-CPS varactor-loaded resonator electrical model, the model of the cross-coupled pair, the model of the buffers and the model of the short-circuit termination, the overall electrical model of the standing wave VCO is as given in Fig. 2.71.

From the model in Fig. 2.71, all the elements of the standing wave VCO can be adjusted, in particular the distribution and the size of the varactors, and of course the length of the quarter-wave resonator in order to achieve on the one hand the frequency of desired oscillation, F_{osc}, and on the other hand the FTR. A compromise exists between the two. Note that a wide FTR involves using the varactors over a wider capacitance variation range, with ultimately a minimum quality factor that is all the lower as the variation range used is high.

The layout of the standing wave VCO is given in Fig. 2.72.

The result of the FTR is given in Fig. 2.73, with a comparison between measurements and post-layout simulations. The measured FTR is equal to 12.1 % around 78.1 GHz centered oscillation frequency. This is much higher as compared to the S-CPS LC-VCO considered in Section 2.6.5.1. However, as mentioned previously, the price to pay is a higher phase noise (-111 dBc/Hz instead of -115 dBc/Hz for the LC-VCO), thus leading to a less good FOM equal to -176 dBc/Hz as compared to -178.6 dBc/Hz for the S-CPS LC-VCO.

2.6.5.3 Conclusion

The two VCOs presented within this section show that slow-wave transmission lines can be successfully used within mm-wave VCOs. The design was illustrated in E-band around 80 GHz, but it is clear that slow-wave transmission lines can be relevant lower in frequency (minimum 50 GHz probably) because they have a shorter length than conventional microstrip lines, and also higher in frequency (maximum probably around 120 GHz), the limitation for the VCOs presented within this section is then essentially linked to the quality factor of the available varactors.

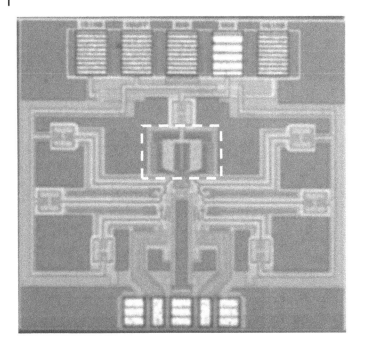

Figure 2.72 Layout of the standing wave VCO (S-CPS resonator highlighted by dashed lines).

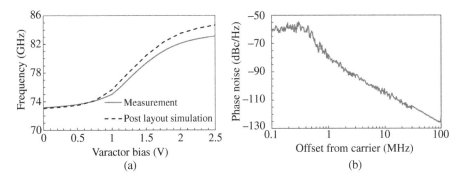

Figure 2.73 Post-layout simulation and measurement of the standing wave VCO of (a) the output frequency and (b) the phase noise.

2.6.6 Phase Shifter

Phase shifters are important devices for phased arrays required in beam-forming and beam-steering applications. Active phase shifters have the great advantage of a compact footprint, but the power consumption can be an issue when mobility or high autonomy applications are targeted. Linearity can also be advantageous for passive phase shifters. Thus, passive phase shifters are relevant if they present low-loss and a reduced size. Their electrical performance is usually compared with the FOM (2.49), where $\Delta\Phi_{max}$ is the maximum phase shift and IL_{max} is the maximum insertion loss.

$$FOM = \frac{\Delta\Phi_{max}}{IL_{max}} \tag{2.49}$$

Various topologies are possible, like, for instance, switched-lines, reflection-type phase shifters, and loaded-lines, but regardless of this choice, a tuning element is mandatory. The first option is to use controllable dielectrics. Ferroelectric materials, such as barium-strontium-titanate (BST), are good candidates up to 10 GHz (Vélu et al., 2007). Then, as the material loss increases with frequency, so does the insertion loss of the device. Contrary to ferroelectrics, liquid crystals (LC) are promising above 10 GHz since the loss tangent is low and decreases with the frequency (Mueller et al., 2005). It is important to note that the switching time of a liquid-crystal-based circuit depends on the liquid crystal layer thickness. The second tuning option is to use switching elements like p-i-n diodes, Metal Oxide Semiconductor Field-Effect Transistor (MOSFETs), or microelectromechanical systems (MEMS) to vary the electrical length by switching between different delay lines or different loads. To get a high phase shift resolution, a large number of loads or delay lines are required, which is surface area consuming. A third possibility is to benefit from the integrated varactor diodes to similarly reach a continuous tuning and a compact device. However silicon varactors exhibit low-quality factors at mm-waves, thus increasing the insertion loss of the phase shifter.

In this section, two solutions to implement a phase shifter with the slow-wave effect are investigated: one with varactors, another one with a combination of liquid crystals and MEMS.

2.6.6.1 Integrated Phase Shifter With Varactors

Beside the slow-wave effect they create, the ribbons in the patterned shield add an extra physical element to a conventional CPW. In this scenario, the designer can use them as a useful part of the design. This was proposed in Bautista et al. (2015) where a center-cut topology was used (see 2.5.1.1) in a S-CPW in combination with a varactor placed at the cut position. This is shown in Fig. 2.74.

This topology allows the tuning of the transmission lines' effective relative dielectric constant. Indeed, the linear capacitance of the varactor-loaded transmission line can be expressed as:

$$C_{lin} = \frac{2C_S C_G (C_v + C_p)}{2C_G(C_v + C_p) + 2C_S(C_v + C_p) + C_S C_G} \tag{2.50}$$

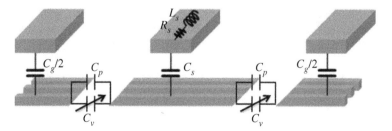

Figure 2.74 Schematic view of the tunable phase shifter combining slow-wave effect and varactors.

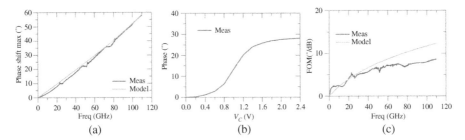

Figure 2.75 Phase shifter measurements versus model results: (a) Maximum phase shift; (b) phase shift as a function of the control voltage at 60 GHz; and (c) FoM.

where C_S represents the signal-ribbon capacitance, C_G the ground-ribbon capacitance, C_p the capacitance between two face-to-face ribbon strips and C_v the variable capacitance of the varactor, respectively. In this scenario, if the capacitance of the varactor can effectively be tuned, so can be the phase shift introduced by the loaded S-CPW.

Using this approach, a phase shifter able to attain a phase shift of 30° at 60 GHz was designed in the 55-nm BiCMOS technology from STMicroelectronics. The measurement results are presented in Fig. 2.75. The proposed phase shifter can operate at a wideband level as its phase shifting capabilities increase with frequency and its FOM remains constant. In addition, this approach allows a miniaturized version of this device as it virtually occupies the same footprint as an unloaded transmission line.

2.6.6.2 Compact Liquid Crystal MEMS Phase Shifter

Another innovative way to realize a mm-wave phase shifter was developed by Franc et al. (2013a). It combines the strong miniaturization of the S-CPW topology with a high tunability thanks to both low-loss liquid crystal technology and MEMS-like behavior.

Device Implementation and Results The schematic view of the phase shifter is presented in Fig. 2.76. A S-CPW is implemented in a CMOS technology. For polarization purposes, grounding vias are added between the CPW ground planes and the shielding ribbons. In addition, anchoring vias are introduced below the ribbons down to the silicon substrate to ensure the ribbons mechanical stability. The silicon dioxide below the CPW signal strip and between the two CPW ground strips is removed thanks to a maskless post-CMOS etching process. Thus, the CPW signal strip is mechanically released (Fig. 2.77), and the empty space is filled with a liquid crystal mixture.

The liquid crystal mixture is dedicated to mm-wave applications and consists of uniaxial anisotropic molecules (Yang and Wu 2006). A bias voltage V_{bias} is applied between two electrodes, namely the CPW signal strip and the shielding ribbons, to rotate the liquid crystal molecules; see Fig. 2.78. When V_{bias} is zero, the electric field is orthogonal to the liquid crystal molecules, and it experiences a medium having the relative effective dielectric constant $\varepsilon_\perp = 2.47$ and a loss tangent $\tan\delta_\perp = 0.015$ (at room temperature and 30 GHz). When the bias voltage exceeds a saturation value, the molecules are completely aligned in parallel to the electric field, and the liquid crystal has the properties $\varepsilon_{//} = 3.2$ and $\tan\delta_{//} = 0.0033$. So, for high bias voltages, the dielectric constant between the two electrodes is maximum, and the capacitance between the CPW signal strip and the shielding ribbons is maximum.

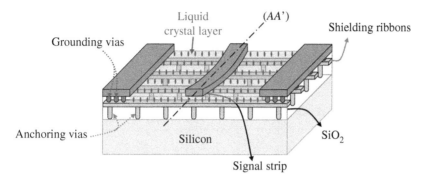

Figure 2.76 Schematic view of the tunable phase shifter combining slow-wave effect, liquid crystals and mechanical bending.

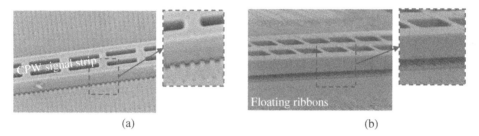

(a)   (b)

Figure 2.77 Photographs of the device: (a) with parallel ribbons embedded in silicon dioxide and (b) after the etching that mechanically released the CPW signal strip.

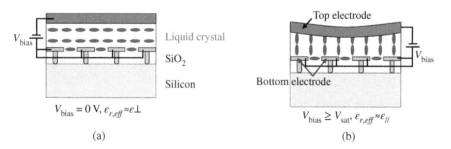

(a)   (b)

Figure 2.78 Schematic view of the (AA') cross-section defined in Fig. 2.76: (a) 0 V bias voltage leading to a perpendicular alignment of the liquid crystal molecules; and (b) high V_{bias} ensuring a parallel alignment of the molecules and a bending of the CPW signal strip due to the electrostatic force.

At the same time, the bias voltage deflects the top electrode (CPW signal strip), introducing a MEMS-like effect that strengthens the capacitive effect by reducing the height between the shielding ribbons and the CPW signal strip.

The measurements exhibit a continuous tuning when the bias voltage increases, Fig. 2.79. The first measurement presents a very high differential phase shift (550° at 45 GHz for a biasing of 20 V). This high phase shift is not reproducible as residual mechanical stress is removed after the first activation. Then a lower but reproducible phase shift of 275° is

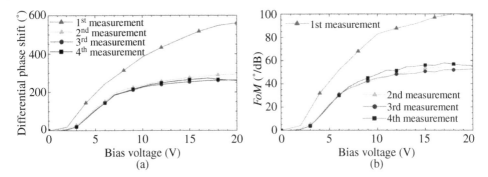

Figure 2.79 Measured differential (a) phase shift and (b) figure of merit (FOM) at 45 GHz versus the applied voltage for the liquid crystal MEMS phase shifter.

achieved at 45 GHz for a 20 V bias voltage. The insertion loss does not vary much with the differential phase shift and equals 5 ± 0.35 dB at 45 GHz, and the 45-GHz FOM reaches $52°/\mathrm{dB}$.

Benefits of the Slow-Wave Effect Here, a comparison of the tunability between the slow-wave and conventional CPW is investigated with the same liquid crystal having a relative dielectric constant varying from $\varepsilon_\perp = 2.47$ to $\varepsilon_{//} = 3.2$, i.e. a 25% tunability. Note that the MEMS deflection has been ignored to remove the multiphysics aspect, thus strongly simplifying the following simulations as Young's modulus and intrinsic stress of the CMOS metallic layers are not provided by the foundry.

For a maximum targeted phase shift $\Delta\Phi_{max}$, the physical length of a liquid crystal filled phase shifter is calculated by (2.51) with f the operating frequency, c_0 the light velocity in vacuum, and $\varepsilon_{r,eff,max}$ ($\varepsilon_{r,eff,min}$, respectively) the maximum (minimum, respectively) effective relative dielectric constant of the liquid crystal filled delay line.

$$l_{\Delta\Phi} = \frac{\Delta\Phi_{max}}{2\pi} = \frac{c_0}{f \cdot \left(\sqrt{\varepsilon_{r,eff,max}} - \sqrt{\varepsilon_{r,eff,min}}\right)} \tag{2.51}$$

Hence, maximizing the factor $\left(\sqrt{\varepsilon_{r,eff,max}} - \sqrt{\varepsilon_{r,eff,min}}\right)$ increases the maximum phase shift for a given length. In the used CMOS technology, and for the considered dimensions of Franc et al. (2013a) and at 45 GHz, this factor equals approximately 0.7 for S-CPW and 0.2 for CPW. It is observed in practice that a $38°/\mathrm{mm}$ differential phase shift is obtained at 45 GHz for the S-CPW delay line, while the conventional CPW counterpart only leads to $11°/\mathrm{mm}$, see Fig. 2.80. So, the slow-wave effect allows reducing the length of the phase shifter by a factor of 3.5 to implement the same phase shift, resulting in better electric performance (insertion loss and FOM).

Conclusion on the Liquid Crystal MEMS Phase Shifter This tunable slow-wave delay line combines the advantages of the liquid crystals and MEMS phase shifters, leading to low insertion loss and high phase shift. In addition, it is a planar structure that also benefits from the compactness of integrated slow-wave transmission lines ($0.38\ \mathrm{mm}^2$). After appropriate packaging, this phase shifter is MMIC compatible, as a low bias voltage of 5 V is enough to realize a $110°$ phase shift.

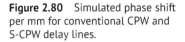
Figure 2.80 Simulated phase shift per mm for conventional CPW and S-CPW delay lines.

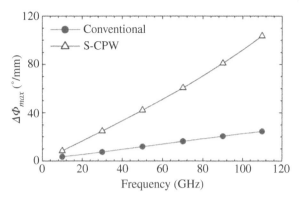

2.6.7 Sensors

So far, the slow-wave concept has been used to design mm-wave circuits. Thanks to the capacitive effect introduced by the shielding ribbons, the S-CPW topology confines the electric field in the dielectric between the two metallic layers, thus offering the prospect of characterizing thin dielectric films at mm-waves.

If the material characterization is targeted on a large frequency band, the use of classical resonators (Gaebler et al., 2008) whose resonance frequency is changed according to the unknown dielectric constant is prohibited. Then, classical wideband characterization techniques rely either on the measurement of a capacitance or on the characteristics of a transmission line (Vo et al., 2008). The former solution is based on a parallel-plate capacitor whose dielectric is an unknown material. It is efficient at low frequencies when the capacitance is directly measured with an LCR meter. When high frequencies are investigated (some tens of GHz), the measurements require a vector network analyzer, and the effect of the access pads has to be removed thanks to a de-embedding step, which reduces the sensitivity. Regarding transmission-lines-based methods, the de-embedding can be avoided thanks to (P. Ferrari et al., 1994). Nevertheless, for thin dielectric films, the sensitivity is usually poor because the surrounding media is required either to ensure a characteristic impedance value not too low (microstrip line case) or to overcome some mechanical constraints (CPW case).

Thanks to the increase in capacitance without changing the inductance, the S-CPW helps keep the characteristic impedance within the correct range, hence allowing a high sensitivity. Based on a CMOS process, a 1-μm thick dielectric film was characterized by Franc et al. (2012a) for various transmission lines, namely microstrip lines, CPW, and S-CPW. The associated sensitivities are plotted in Fig. 2.81, which compares the extracted parameters (relative effective dielectric constant, attenuation constant) as a function of the unknown dielectric properties (relative dielectric constant and loss tangent).

The method with the best sensitivity is the one exhibiting the highest slope. Then the S-CPW reaches a better precision as the dynamic for the relative dielectric constant is 3.8 times higher compared to conventional microstrip lines and CPW. The same conclusion is obtained with the dielectric loss tangent, whose dynamic is increased by a factor of 1.9 times with the use of the S-CPW.

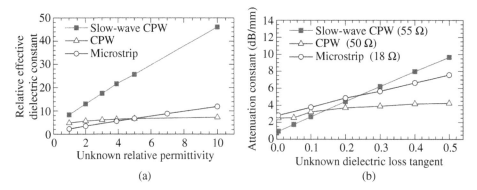

Figure 2.81 Simulation of the 60-GHz extracted parameters versus the unknown dielectric properties for the proposed method and classical ones: (a) Relative effective dielectric constant versus unknown relative dielectric constant; (b) Attenuation constant versus unknown dielectric loss tangent.

This sensor principle can also be applied to very low volumes of liquids: a post-CMOS process is then required to create a cavity between the two conductors (CPW signal strip and shielding ribbons) that is finally filled with the liquid under-test, (Franc et al., 2012c).

2.7 Conclusion

In this chapter, various integrated slow-wave transmission lines have been presented, namely S-CPW, S-CPS, and coupled S-CPW. The patterned shield introduces several new parameters offering high flexibility. Then, it becomes crucial to derive some design rules to help the designer benefit best from the new degrees of freedom in a given technology. For each topology, the electrical model was derived based on its electromagnetic behavior. Those models can then be included in the design process to avoid time- and memory-consuming full-wave simulations.

To fully show the potentiality of slow-wave coplanar transmission lines in integrated technologies, many circuits, benefiting from the slow-wave effect, were developed and measured. They highlight that any mm-wave frequency can take advantage of the topology, especially to reduce the footprint area, increase the quality factor, and add design flexibility.

Finally, it is important to note that the performance of slow-wave coplanar transmission lines remains better than the conventional extensively used microstrip lines only below 150 GHz.

References

Abdel Aziz, M., Issa, H., Kaddour, D., Podevin, F., Safwat, A. M. E., Pistono, E., Duchamp, J.-M., Vilcot, A., Fournier, J.-M., & Ferrari, P. (2009). Shielded coplanar striplines for RF integrated applications. *Microwave and Optical Technology Letters*, 51(2), 352–358. https://doi.org/10.1002/mop

Ahn, H. R. (2011). Modified asymmetric impedance transformers (MCCTs and MCVTs) and their application to impedance-transforming three-port 3-dB power dividers. *IEEE Transactions on Microwave Theory and Techniques*, 59(12 PART 2), 3312–3321. https://doi .org/10.1109/TMTT.2011.2171708

Andress, W. F., & Ham, D. (2005). Standing wave oscillators utilizing wave-adaptive tapered transmission lines. *IEEE Journal of Solid-State Circuits*, 40(3), 638–651. https://doi.org/10 .1109/JSSC.2005.843600

Antonini, G., Orlandi, A., & Paul, C. R. (1999). Internal impedance of conductors of rectangular cross section. *IEEE Transactions on Microwave Theory and Techniques*, 40(7), 979–985. https://doi.org/10.1109/22.775429

Bastida, E. M., & Donzelli, G. P. (1979). Periodic slow-wave low-loss structures for monolithic GaAs microwave integrated circuits. *Electronics Letters*, 15(19), 581–582.

Bautista, A., Franc, A. -L., & Ferrari, P. (2015). Accurate parametric electrical model for slow-wave CPW and application to circuits design. *IEEE Transactions on Microwave Theory and Techniques*, 63(12), 4225–4235. https://doi.org/10.1109/TMTT.2015.2495242

Burdin, F., Podevin, F., & Ferrari, P. (2016). Flexible and miniaturized power divider. *International Journal of Microwave and Wireless Technologies*, 8(3), 547–557.

Cathelin, A., Martineau, B., Seller, N., Douyère, S., Gorisse, J., Pruvost, S., Raynaud, C., Gianesello, F., Montusclat, S., Voinigescu, S. P., Niknejad, A. M., Belot, D., & Schoellkopf, J. P. (2007). Design for Millimeter-wave Applications in Silicon Technologies. *European Solid State Circuits Conference*, 1, 464–471.

Cédrat. (n.d.). Flux. In *[Online], Available:* https://cedrat.com.

Cheung, T. S. D., & Long, J. R. (2006). Shielded passive devices for silicon-based monolithic microwave and millimeter-wave integrated circuits. *IEEE Journal of Solid-State Circuits*, 41(5), 1183–1200. https://doi.org/10.1109/JSSC.2006.872737

Chirala, M. K., & Floyd, B. A. (2006). Millimeter-wave lange and ring-hybrid couplers in a silicon taechnology for E-Band applications. *International Microwave Symposium (IMS)*, 1547–1550.

Engen, G. F., & Hoer, C. A. (1979). 'Thru-Reflect-Line': an improved technique for calibrating the dual six-port automatic network analyzer. *IEEE Transactions on Microwave Theory and Techniques*, 27(12), 987–993.

Ferrari, P., Flechet, B., & Angenieux, G. (1994). Time domain characterization of lossy arbitrary characteristic impedance transmission lines. *IEEE Microwave and Guided Wave Letters*, 4(6), 177–179. https://doi.org/10.1109/75.294284

Franc, A.-L., Ferrari, P., & Rehder, G. (2012c). A millimeter-wave integrated sensor for the dielectric constant characterization of pico-liter volumes of liquids. *International Semiconductor Conference Dresden-Grenoble*, 1–4.

Franc, A. L., Karabey, O. H., Rehder, G., Pistono, E., Jakoby, R., & Ferrari, P. (2013a). Compact and broadband millimeter-wave electrically tunable phase shifter combining slow-wave effect with liquid crystal technology. *IEEE Transactions on Microwave Theory and Techniques*, 61(11), 3905–3915. https://doi.org/10.1109/TMTT.2013.2282288

Franc, A.-L., Pistono, E., Corrao, N., Gloria, D., & Ferrari, P. (2011). Compact high-Q, low-loss transmission lines and power splitters in RF CMOS technology. *2011 IEEE MTT-S International Microwave Symposium, Baltimore, MD*, 1–4.

Franc, A.-L., Pistono, E., & Ferrari, P. (2010). Design guidelines for high performance slow-wave transmission lines with optimized floating shield dimensions. *European Microwave Conference (EuMC)*, 1190–1193.

Franc, A. L., Pistono, E., & Ferrari, P. (2012a). Characterization of thin dielectric films up to mm-wave frequencies using patterned shielded coplanar waveguides. *IEEE Microwave and Wireless Components Letters*, 22(2), 100–102. https://doi.org/10.1109/LMWC.2011.2180517

Franc, A.-L., Pistono, E., & Ferrari, P. (2015). Dispersive Model for the Phase Velocity of Slow-Wave CMOS Coplanar Waveguides. *2015 45th European Microwave Conference: 7-10 September 2015, Paris, France*, 48–51.

Franc, A. L., Pistono, E., Gloria, D., & Ferrari, P. (2012b). High-performance shielded coplanar waveguides for the design of CMOS 60-GHz bandpass filters. *IEEE Transactions on Electron Devices*, 59(5), 1219–1226. https://doi.org/10.1109/TED.2012.2186301

Franc, A. L., Pistono, E., Meunier, G., Gloria, D., & Ferrari, P. (2013b). A lossy circuit model based on physical interpretation for integrated shielded slow-wave CMOS coplanar waveguide structures. *IEEE Transactions on Microwave Theory and Techniques*, 61(2), 754–763. https://doi.org/10.1109/TMTT.2012.2231430

Gaebler, A., Goelden, F., Mueller, S., & Jakoby, R. (2008). Triple-mode cavity perturbation method for the characterization of anisotropic media. *European Microwave Conference*, 909–912.

Gianesello, F., Gloria, D., Montusclat, S., Raynaud, C., Boret, S., Dambrine, G., Lepilliet, S., Martineau, B., & Pilard, R. (2007). 1.8 dB insertion loss 200 GHz CPW band pass filter integrated in HR SOI CMOS technology. *IEEE/MTT-S International Microwave Symposium*, 453–456.

Gomes, L., Sharma, E., Souza, A. A. L., Serrano, A. L. C., Rheder, G. P., Pistono, E., Ferrari, P., & Bourdel, S. (2021). 77.3-GHz standing-wave oscillator based on an asymmetrical tunable slow-wave coplanar stripline resonator. *IEEE Transactions on Circuits and Systems I: Regular Papers*, 68(8), 3158–3169. https://doi.org/10.1109/TCSI.2021.3060579

Hammerstad, E. (1981). Computer-aided design of microstrip couplers with accurate discontinuity models. *IEEE MTT-S International Microwave Symposium Digest (IMS)*, 54–56. https://doi.org/10.1109/MWSYM.1981.1129818

Hao, Z. C., & Hong, J. S. (2008). Ultra-wideband bandpass filter using multilayer liquid-crystal-polymer technology. *IEEE Transactions on Microwave Theory and Techniques*, 56(9), 2095–2100. https://doi.org/10.1109/TMTT.2008.2002228

Hasegawa, H., Furukawa, M., & Yanai, H. (1971). Properties of microstrip line on Si-SiO2 system. *IEEE Transactions on Microwave Theory and Techniques*, 19(11), 869–881.

Hirota, T., Minakawa, A., & Muraguchi, M. (1990). Reduced-size branch-line and rat-race hybrids for Uniplanar MMIC's. *IEEE Transactions on Microwave Theory and Techniques*, 38(3), 270–275.

Hong, J., & Lancaster, M. J. (2001). *Microstrip Filters for RF/Microwave Applications* (Vol. 7). John Wiley & Sons.

Horst, S., Bairavasubramanian, R., Tentzeris, M. M., & Papapolymerou, J. (2007). Modified Wilkinson power dividers for millimeter-wave integrated circuits. *IEEE Transactions on Microwave Theory and Techniques*, 55(11), 2439–2446. https://doi.org/10.1109/TMTT.2007 .908672

Kaddour, D., Issa, H., Franc, A.-L., Corrao, N., Pistono, E., Podevin, F., Fournier, J.-M., Duchamp, J.-M., & Ferrari, P. (2009). High-*Q* slow-wave coplanar transmission lines on 0.35 μm CMOS process. *IEEE Microwave and Wireless Components Letters*, 19(9), 542–544. https://doi.org/10.1109/LMWC.2009.2027053

Kirschning, M., & Jansen, R. H. (1982). Accurate model for effective dielectric constant of microstrip with validity up to millimetre-wave frequencies. *Electronics Letters*, 18(6), 272–273. https://doi.org/10.1049/el:19820186

Koolen, M. C. A. M., Geelen, J. A. M., & Versleijen, M. P. J. G. (1991). An improved deembedding technique for on-wafer high-frequency characterization. *IEEE Bipolar Circuits and Technology Meeting*, 188–191.

Lai, I. C. H., & Fujishima, M. (2007). High-*Q* slow-wave transmission line for chip area reduction on advanced CMOS processes. *IEEE International Conference on Microelectronic Test Structures (ICMTS)*, 192–195. https://doi.org/10.1109/ICMTS.2007.374481

Lee, J. J., & Park, C. S. (2010). A slow-wave microstrip line with a high-*Q* and a high dielectric constant for millimeter-wave CMOS application. *IEEE Microwave and Wireless Components Letters*, 20(7), 381–383. https://doi.org/10.1109/LMWC.2010.2049430

Lu, T. Y., Yu, C. Y., Chen, W. Z., & Wu, C. Y. (2013). Wide tunning range 60 GHz VCO and 40 GHz DCO using single variable inductor. *IEEE Transactions on Circuits and Systems I: Regular Papers*, 60(2), 257–267. https://doi.org/10.1109/TCSI.2012.2215795

Lugo-alvarez, J., Bautista, A., Podevin, F., & Ferrari, P. (2014). High-directivity compact Slow-wave CoPlanar Waveguide couplers for millimeter-wave applications. *2014 44th European Microwave Conference, Rome, Italy*, 1072–1075.

Ma, Y., Rejaei, B., & Zhuang, Y. (2008). Artificial dielectric shields for integrated transmission lines. *Microwave and Wireless Components Letters, IEEE*, 18(7), 431–433.

Mandal, M. K., & Sanyal, S. (2007). Reduced-length rat-race couplers. *IEEE Transactions on Microwave Theory and Techniques*, 55(12), 2593–2598. https://doi.org/10.1109/TMTT.2007.910058

Mangan, A. M., Voinigescu, S. P., Ming-Ta, Y., & Tazlauanu, M. (2006). De-embedding transmission line measurements for accurate modeling of IC designs. *IEEE Transactions on Electron Devices*, 53(2), 235–241. https://doi.org/10.1109/TED.2005.861726

Margalef-Rovira, M. et al. (2020). Design of mm-Wave Slow-Wave-Coupled Coplanar Waveguides. *IEEE Transactions on Microwave Theory and Techniques*, 68(12), 5014–5028. https://doi.org/10.1109/TMTT.2020.3015974

Morandini, Y., Gianesello, F., Boret, S., Lasserre, S., Gloria, D., & Pekarik, J. (2010). Evaluation of sub-32nm CMOS technology for Millimeter wave applications. *European Microwave Conference (EuMC)*, 417–420.

Mueller, S., Penirschke, A., Damm, C., Scheele, P., Wittek, M., Weil, C., & Jakoby, R. (2005). Broad-band microwave characterization of liquid crystals using a temperature-controlled coaxial transmission line. *IEEE Transactions on Microwave Theory and Techniques*, 53(6 II), 1937–1945. https://doi.org/10.1109/TMTT.2005.848842

Oliver, B. M. (1954). Directional electromagnetic couplers. *Proceedings of the IRE*, 42(11), 1686–1692.

Quémerais, T., Moquillon, L., Fournier, J. M., & Benech, P. (2010). 65-, 45-, and 32-nm aluminium and copper transmission-line model at millimeter-wave frequencies. *IEEE*

Transactions on Microwave Theory and Techniques, 58(9), 2426–2433. https://doi.org/10.1109/TMTT.2010.2058277

Quendo, C., Rius, E., & Person, C. (2003). Narrow bandpass filters using dual-behavior resonators. *IEEE Transactions on Microwave Theory and Techniques*, 51(3), 734–743. https://doi.org/10.1109/TMTT.2003.808729

Sharma, E., Saadi, A. A., Margalef-Rovira, M., Pistono, E., Barragan, M. J., Lisboa De Souza, A. A., Ferrari, P., & Bourdel, S. (2020). Design of a 77-GHz LC-VCO with a slow-wave coplanar stripline-based inductor. *IEEE Transactions on Circuits and Systems I: Regular Papers*, 67(2), 378–388. https://doi.org/10.1109/TCSI.2019.2926415

Suh, Y.-H., & Chang, K. (2002). Coplanar stripline resonators modeling and applications to filters. *IEEE Transactions on Microwave Theory And Techniques*, 50(5), 1289–1296.

Tang, X. L., Franc, A.-L., Pistono, E., Siligaris, A., Vincent, P., Ferrari, P., & Fournier, J.-M. (2012). Performance improvement versus CPW and loss distribution analysis of S-CPW in 65 nm HR-SOI CMOS technology. *IEEE Transactions on Electron Devices*, 59(5), 1279–1285. https://doi.org/10.1109/TED.2012.2186969

Tentzeris, M. M., Laskar, J., Papapolymerou, J., Pinel, S., Palazzari, V., Li, R., DeJean, G., Papageorgiou, N., Thompson, D., Bairavasubramanian, R., Sarkar, S., & Lee, J. H. (2004). 3-D-Integrated RF and millimeter-wave functions and modules using liquid crystal polymer (LCP) system-on-package technology. *IEEE Transactions on Advanced Packaging*, 27(2), 332–340. https://doi.org/10.1109/TADVP.2004.828814

Vecchi, F., Repossi, M., Eyssa, W., Arcioni, P., & Svelto, F. (2009). Design of low-loss transmission lines in scaled CMOS by accurate electromagnetic simulations. *IEEE Journal of Solid-State Circuits*, 44(9), 2605–2615. https://doi.org/10.1109/JSSC.2009.2023277

Vélu, G., Blary, K., Burgnies, L., Marteau, A., Houzet, G., Lippens, D., & Carru, J. C. (2007). A 360° BST phase shifter with moderate bias voltage at 30 GHz. *IEEE Transactions on Microwave Theory and Techniques*, 55(2), 438–443. https://doi.org/10.1109/TMTT.2006.889319

Vo, T. T., Lacrevaz, T., Bermond, C., Bertaud, T., Fléchet, B., Farcy, A., Morand, Y., Blonkowski, S., Torres, J., Guigues, B., & Defaÿ, E. (2008). In situ microwave characterisation of medium-k HfO2 and high-k SrTiO3 dielectrics for metal-insulator-metal capacitors integrated in back-end of line of integrated circuits. *IET Microwaves, Antennas and Propagation*, 2(8), 781–788. https://doi.org/10.1049/iet-map:20070344

Wilkinson, E. J. (1960). An N-Way Hybrid Power Divider. *IRE Transactions on Microwave Theory and Techniques*, 8(1), 116–118.

Yang, D.-K., & Wu, S.-T. (2006). *Fundamentals of Liquid Crystal Devices*. Wiley.

Zhong, G., & Koh, C.-K. (2003). Exact closed-form formula for partial mutual inductances of rectangular conductors. *IEEE Transactions on Circuits and Systems I: Fundamental Theory and Applications*, 50(10), 1349–1352. https://doi.org/10.1109/TCSI.2003.817778

3

Slow-Wave Microstrip Lines

Hamza Issa[1] and Ariana Lacorte Caniato Serrano[2]

[1]*Faculty of Engineering, Beirut Arab University, Beirut, Lebanon*
[2]*Polytechnic School, University of São Paulo, São Paulo, Brazil*

3.1 Introduction

Slow-wave microstrip lines (S-MS) have been demonstrated for decades from microwave to millimeter wave frequencies in different technologies, such as integrated lines on silicon (Gammand & Bajon, 1990; Lee & Park, 2010; Wang et al., 2009), and printed circuit board (PCB) (Yang et al., 1998; Coulombe et al., 2007); and different configurations, using patterned ground shield transmission lines (Lee & Park, 2010; Yang et al., 1998), transmission lines loaded with distributed elements (Gammand & Bajon, 1990), using transmission lines with different characteristic impedances to create the slow-wave effect (Wang et al., 2009), using metallic vias (Coulombe et al., 2007), among others. S-MS lines were employed in the last decades for compact devices, such as directional coupler in PCB (Luong et al., 2019), branch-line couplers in PCB (Abouchahine et al., 2018), meandered branch-line coupler in CMOS (Acri et al., 2019), and bandpass filters (Wu et al., 2002; Li et al., 2005), (Evans et al., 2012). As it holds true for classic microstrip lines, the S-MS lines are simple to design, fabricate, and integrate.

In this chapter, we consider a specific type of S-MS lines based on the "bed of nails" concept. This type of S-MS line is advantageous as it can be constructed using simple blind vias connected to a solid ground plane. Additionally, they can be modeled using a distributed model, greatly accelerating the design process time.

This chapter is organized as follows: First, the slow-wave effect of the S-MS lines is presented in Section 3.2 with its topology, followed by how the concept is physically structured in different technologies at low microwave frequencies, in Section 3.3, and at millimeter waves in Section 3.4. A physics-based model of the S-MS lines is shown and validated with experimental results in Section 3.5. Different applications are given in 3.6 with design and measurement of devices based on the S-MS lines and the chapter closes in 3.7 with a discussion on the design of S-MS lines for integrated circuits in CMOS technologies.

Slow-Wave Microwave and mm-Wave Passive Circuits, First Edition. Edited by Philippe Ferrari,
Anne-Laure Franc, Marc Margalef-Rovira, Gustavo P. Rehder, and Ariana Lacorte Caniato Serrano.
© 2025 John Wiley & Sons Ltd. Published 2025 by John Wiley & Sons Ltd.

3.2 Principle of Slow-Wave Microstrip Lines

The principle of the S-MS line is to increase the capacitance of a classical microstrip line significantly more than the changes in its inductance. The simplest and most commonly used S-MS line topology is presented in Fig. 3.1. This topology can operate in a large range of frequencies up to millimeter waves using different technologies to achieve the appropriate ratio of dimensions that causes the slow-wave effect. In this chapter, three important technologies are discussed for S-MS lines: the standard multi-layered PCB technology with metallic vias, dedicated to radiofrequency applications (Coulombe et al., 2007; Machac, 2006); the metallic nanowire membrane (MnM) technology with metallic nanowires, dedicated to millimeter waves (Serrano et al., 2014a, 2014b); and finally, the concept is also presented in CMOS technology for integrated circuits. The difference between these technologies is the size of the structures, the layers used, and the frequency band where the S-MS lines can operate.

In all cases, to achieved the slow-wave effect, a classical microstrip line with dielectric layer D_1 is used and one or more layers are added between this dielectric layer and the ground plane. An array of blind metallic vias or nanowires, forming a bed of nails, is connected to the ground plane in the dielectric layer above it, D_2 (dashed volume in Fig. 3.1). The array does not cross D_1 and, therefore, does not touch the signal strip. The slow-wave effect is obtained as the electric field is concentrated in the upper dielectric layer D_1, which is immediately on the top of the end of the "nails". The magnetic field remains as in classical microstrip lines flowing through the entire volume formed by D_1 and D_2. In this manner, electric and magnetic fields are physically separated in volume D_2, where the electric field can be almost null depending on the density of vias/nanowires. This also means that the electrical characteristics of the dielectric to be considered in the design of a S-MS line is D_1 only (the dielectric layer(s) between the signal strip and the bed of nails), as it will be shown in the following sections.

This physical behavior is illustrated in Fig. 3.2, comparing the electric and magnetic field patterns of a classical microstrip line in (a), and of a S-MS line in (b), both with the same width W in a transversal view. The fields were simulated in the 3D electromagnetic (EM) simulator Ansys Electronics.

The change in the electric field leads to an increase in the linear capacitance C, while the linear inductance L is almost the same as compared to classical microstrip lines. Therefore, the phase velocity is reduced expressed as equation (3.1) and the effective dielectric constant $\varepsilon_{r_{eff}}$ expressed as equation (3.2) is increased. In these equations, c_0 is the light speed

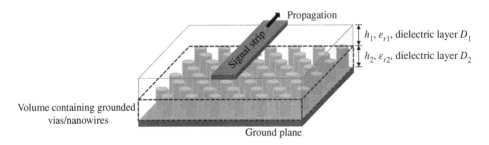

Figure 3.1 Slow-wave microstrip (S-MS) line based on the "bed of nail" concept.

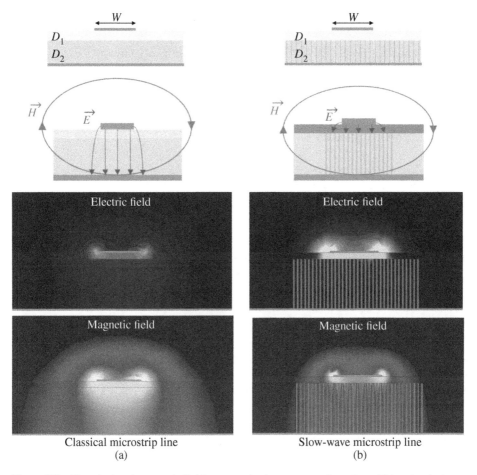

Figure 3.2 Electrical and magnetic field patterns in the transversal section of (a) a classical microstrip line and (b) a S-MS line.

in vacuum. This means that, for the same desired transmission line electrical length θ, the signal strip can be made physically shorter. From equation (3.3), increasing the value of $\varepsilon_{r_{eff}}$ leads to decreasing the guided wavelength λ_g. This in turn decreases the phase constant β in equation (3.4). Therefore, from equation (3.5), since the phase constant β decreases, the physical length l has to be made shorter in order to keep a constant electrical length θ.

$$v_\varphi = 1/\sqrt{LC} \tag{3.1}$$

$$\varepsilon_{r_{eff}} = c_0\sqrt{LC} \tag{3.2}$$

$$\lambda_g = \frac{v_\varphi}{f} = \frac{c_0}{\sqrt{\varepsilon_{r_{eff}}} \cdot f} \tag{3.3}$$

$$\beta = \frac{2\pi}{\lambda_g} \tag{3.4}$$

$$\theta = \beta l \tag{3.5}$$

3.3 PCB Technology

At lower microwave frequencies, PCB technology is commonly employed for building microwave circuits. Its straight forward fabrication process, seamless integration with other circuits, and cost-effectiveness contribute to its widespread popularity across various devices for the consumer market applications. In this section, S-MS lines in PCB technology are considered and briefly described. Both single lines and coupled lines are considered, and the topology of their electrical model is presented.

3.3.1 Slow-Wave Microstrip Line

In PCB technology, the slow-wave phenomenon can be created by inserting metallic blind vias underneath the dielectric layer holding the signal strip. The vias are connected to the ground plane from one side and remain floating vertically in the substrate on the other side. The complete structure can be created using a stack of different substrates. An example of a multi-layered PCB technology for S-MS lines is presented in Fig. 3.3. The lower dielectric layer in which the vias are placed should be relatively rigid to keep the pattern of the vertical vias. On the other hand, the intermediate dielectric layer should have flexible properties in order to ensure good adhesion between the upper and lower dielectric layers. During the adhering process, the height of the intermediate substrate is subjected to a maximum decrease of 40%. This decrease affects the design characteristics and must be considered while designing. For instance, in S-MS lines design, the decrease in the height of the intermediate substrate can cause a decrease in the characteristic impedance and the phase velocity values by 10–15%.

The vias with diameter d are separated by a center-to-center distance S; d/d_{pad} are the via/pads diameters, and e/e_{pad} the via/pads pitches. Due to technology constraints, a metal round pad is formed above the vias in the intermediate layer with diameter d_{pad}, whereas

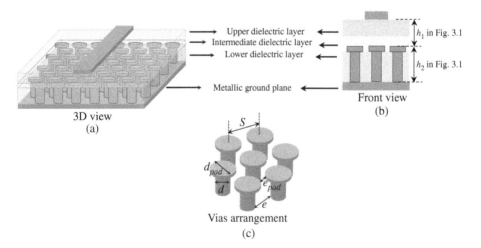

Figure 3.3 Schematic of a S-MS line in printed circuit board (PCB) technology using vias with pads on their top in (a) 3D view and (b) front view. (c) Schematic 3D view of the metallic vias organization with pads.

Figure 3.4 Schematic of a coupled S-MS line in PCB technology.

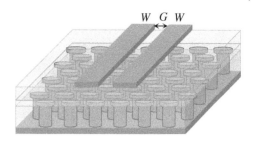

$W \quad G \quad W$

the vias themselves are placed inside the lower dielectric layer. The minimum dimensions for d, d_{pad} and S are imposed by the technology. For the calculi to be realized in the next sections, the height of the dielectric layer related to h_1 in Fig. 3.1 includes the intermediate and upper dielectric layers, i.e. between the top of the vias pads and the bottom of the signal strip, as indicated in Fig. 3.3. The height of the dielectric layer related to h_2 in Fig. 3.1 is considered as the layer where the vias and the pads are, thus including the lower and part of the intermediate dielectric layers.

In order to guarantee the efficiency of the slow-wave effect, the distance e_{pad} between two adjacent pads must be smaller than the overall thickness h_1. In this case, only a minimum amount of electrical field can penetrate the volume where the vias are placed.

3.3.2 Slow-Wave Coupled Lines

The same concept can be applied to the design of slow-wave coupled lines. The achived impedances and couplings differ from those of classical coupled lines making them well-suited for applications such as directional couplers. Figure 3.4 depicts the structure of a coupled S-MS line on the same PCB substrate used for the single S-MS line described in Fig. 3.3.

The coupled S-MS is composed of two metallic strips of width W, separated by a spacing G, patterned on the top metal layer of the PCB substrate, the same as in Fig. 3.3. This structure, like classical coupled lines, supports two distinct propagation modes: the even-mode and the odd-mode.

In Section 3.2, it was explained that single S-MS lines structure naturally exhibits significant confinement of the electrical field within the upper dielectric layers, extending up to the top of the via pads, leading to an elevated capacitance compared to classical microstrip lines. The same effect is present in the coupled S-MS structure, which leads to elevated odd- and even-mode capacitances and the values of these capacitances are relatively closer to each other when compared to classical microstrip coupled lines. However, the inductances of the odd- and even-modes are almost those of classical microstrip coupled lines and can be quite different from each other. This implies a significant disparity in the phase velocities of the odd- and even-modes, which will be discussed further along with its electrical model.

3.4 Metallic Nanowire Membrane Technology

At higher microwave frequencies and millimeter waves, PCB technology is no longer suitable due to the limitations of its fabrication process. To overcome these constraints, Metallic

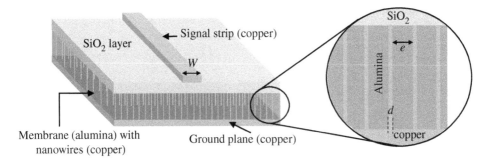

Figure 3.5 Illustration of a S-MS line on the metallic nanowire-filled membrane (MnM)-substrate. The inset shows the dimensions of the diameter *d* of the nanowire and the spacing between the nanowires e.

nanowire Membrane (MnM) technology as emerged as an attractive solution for slow-wave devices operating at such frequencies. Fundamentally, the metallic nanowire-filed membrane (MnM) technology is similar to the technologies used in PCB (Coulombe et al., 2007) or plastic (Dubuc et al., 2009b; Dubuc et al., 2009a). However, the drawback of the PCB technologies is the electrically large vias spaced too far apart, which fail to totally confine the electric field in the top dielectric layer, thus reducing the slow-wave effect and increasing propagation loss.

The MnM S-MS line, shown in Fig. 3.5, is composed of two dielectric layers. The bottom layer is made of a 50-μm-thick InRedox Anodic Aluminum Oxide (AAO), a porous alumina membrane where metallic nanowires are grown inside the nanopores and connected to a metallic layer, the ground plane. This membrane has pore area density (*dst*). The membrane with nanowires is covered by a dielectric layer, which in this case is silicon dioxide (SiO_2). There are two copper layers, one at the bottom, used as ground plane, and also the base layer for nanowire growth. The top copper layer, on top of those two dielectrics stack, is used for the microstrip line signal strip. In practical realizations, the membrane can have a thickness between 50 and 100 μm, and the dielectric layer between 0.5 to 3 μm, depending on the slow-wave factor (SWF) and characteristic impedances range that are targeted. The nanowires are thin enough (nearly four times smaller than the skin depth at 100 GHz) to allow operation with good performance well above 100 GHz with very moderate eddy current losses, and low conductive losses, since the equivalent resistance is reduced by the high density of parallel nanowires.

The porous alumina membrane is fabricated through electrochemical oxidation of aluminum at specific anodization voltages, as described in Masuda and Fukuda (1995). During anodization, various nanopores with different sizes (pore diameter *d* and interpore distance *e*) are formed. Figure 3.6 (a) depicts a top view and (b) a bottom view of the pores in the membrane, and (c) shows the profile view. Different acid solutions, such as sulfuric acid, phosphoric acid or oxalic acid, can be used in the formation process, with each acid solution requiring a specific electric potential range. In this way, this technology is flexible as the geometric parameters can be adjusted during substrate fabrication based on the chosen solution and electric potential, and there is no fixed size limit for the sample.

The electrical characteristics of the membrane are relative dielectric constant $\varepsilon_r = 6.7$ and loss tangent $\tan \delta - 0.01$, up to at least 100 GHz.

Top side   Bottom side   Profile view

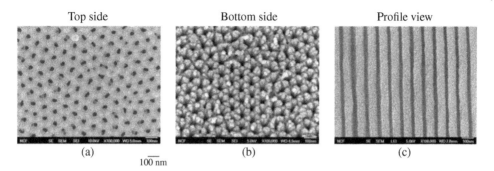

(a)   (b)   (c)

$\overline{100\,\text{nm}}$

Figure 3.6 Nanopores in a membrane with 40 nm pore diameter: (a) top side, and (b) bottom side, and (c) profile view.

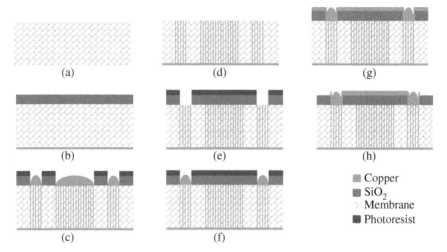

(a)   (d)   (g)

(b)   (e)   (h)

                                                        ■ Copper
                                                        ■ SiO_2
                                                        Membrane
                                                        ■ Photoresist

(c)   (f)

Figure 3.7 Steps of the fabrication process for S-MS lines in the MnM technology: (a) membrane, (b) SiO_2 deposition, (c) photoresist, (d) mask removal and polishing, (e) dielectric layer development, (f) copper layer, (g) photoresist removal, and (h) top copper layer deposition.

The porous alumina membrane utilized in the MnM technology serves as a template for the growth of copper nanowires within the pores, distinguishing it from other millimeter-wave substrates such as low-temperature co-fired ceramic (LTCC), glass, silicon, liquid crystal polymer (LCP), and others. The size of the pores in the membrane can vary from 10 to 160 nm, but optimum results are typically achieved with a pore diameter of approximately 55 nm. Figure 3.7 illustrates the various steps involved in the fabrication process of the MnM technology, from (a) to (h).

A quick view of its fabrication process is given here regarding Figure 3.7. Initially, the membrane is cleaned (a) in boiling trichloroethylene, followed by acetone and isopropyl alcohol.

Then, a masking SiO_2 of about 330 nm is deposited (b) using RF sputtering. This is followed by the ground copper deposition, when a Ti seed is deposited for adherence using RF sputtering, followed by copper seed deposition using RF sputtering. The copper seed layer is then thickened by electrodeposition, reaching a copper thickness of 2–3 μm.

In (c), the photoresist – the SiO$_2$ mask layer – is coated and then exposed and developed. The photoresist is opened in a buffered oxide etchant (BOE) solution. The photoresist is left as an additional protection layer on the SiO$_2$ mask for the nanowire growth. The nanowires are grown using a periodic pulsed current. The positive current density is adjusted to each mask employed, while the negative current density is approximately 3–4 times greater than the positive current.

In (d), the SiO$_2$ mask is removed with acetone and acetic acid, and then the membrane is polished: a rough polishing to remove the excess of copper, followed by a fine polishing to remove the roughness. Deionized water is used to clean the membrane in this step.

A final SiO$_2$ layer is deposited (e) for the dielectric layer of the final S-MS lines as described in step (b) and coated, exposed/developed, and opened as described in (c). In this step, the opening exposes some nanowire regions for contact with the ground, when needed, for example, for the ground connection of the pads of the structures. As in step (c), the photoresist is kept for protection in the next step. Using alignment structures, the mask can be aligned.

In (f), copper is electrodeposited to establish an electrical contact between the exposed nanowire areas and the upper copper layer, which will be deposited in step (g).

In step (g), the photoresist is removed, and a copper seed layer, utilizing titanium for adherence of the top layer, is deposited. This layer is then thickened as in (c).

Finally, the strips are defined (h) in the top copper layer (or even in the ground plane, when needed). Patterns are defined using a photoresist mask and etched. Afterwards, the titanium is removed.

3.5 Electrical Model

This section deals with developing a distributed electric model of S-MS transmission lines using metallic vias. Two segments of the proposed distributed model are represented in

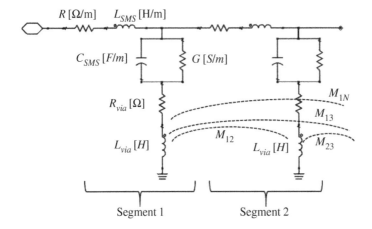

Figure 3.8 Topology of the distributed model of the slow-wave microstrip line.

Fig. 3.8 which is based on a classical transmission line *RLGC* distributed-element model, where R, L, G, and C are the resistance, inductance, conductance and capacitance per meter of a transmission line and which has been adapted to take the effect of the metallic vias into account, in particular the magnetic coupling between vias. R, L_{SMS}, G, and C_{SMS} are the distributed elements of a classical microstrip line model. Each metallic via is modeled by the inductance L_{via} and the resistance R_{via}. In each segment, the mutual inductance M_{ij} models the mutual coupling between the L_{via} of this segment and of all other segments, with *i* and *j* the different segments of the model.

It is noteworthy that, in addition to incorporating R_{via}, L_{via} and M_{ij} into the classical *RLGC* model, the linear capacitance C_{SMS} and the linear inductance L_{SMS} of the S-MS line are calculated in a different manner from classical microstrip lines. In Fig. 3.2(b), it is illustrated that the electric field predominantly exists on the upper dielectric layer D_1, partially penetrating into the dielectric layer below it, between the vias. This penetration depends on the via/pad pitch and directly affects C_{SMS}. Despite the small effect of the metallic vias in L_{SMS}, it exists and changes the way this linear inductance is calculated.

The initial consideration for this analysis is the area density of metallic vias to be accounted for in the calculations of C_{SMS} and L_{SMS} for each technology. In specific technologies like PCB, there are pads situated on top of the vias, as depicted in Fig. 3.3. These pads are integral to the slow-wave effect by increasing the metal density from an electrical field perspective, thereby increasing C_{SMS}.

The density of these pads dst_{pad} is expressed as:

$$dst_{pad} = \frac{2\pi}{\sqrt{3}} \frac{r_{pad}^2}{(2r_{pad} + e_{pad})^2} \tag{3.6}$$

where r_{pad} is the radius of the pads with $r_{pad} = \frac{d_{pad}}{2}$, and e_{pad} is the pad pitch, as illustrated in Fig. 3.3, which shows the 3D view of an area of vias with pads. The area density of the vias can be defined in the same way as for equation (3.6) as:

$$dst_{via} = \frac{2\pi}{\sqrt{3}} \frac{r^2}{(2r + e)^2} \tag{3.7}$$

where *r* is the radius of each via beneath the pad with $r = \frac{d}{2}$ and *e* is the via pitch, also shown in Fig. 3.3. If the technology, such as the MnM technology, has no pads on top of the vias, $dst_{pad} = dst_{via}$.

3.5.1 Linear Capacitance C_{SMS}

The electrical field lines in an S-MS line with a width *W* are depicted in the transverse view in Fig. 3.9. Typically, the upper dielectric layer is relatively thin compared to *W*, i.e., $W/h_1 > 1$, aiming to enhance the slow-wave effect. In this way, the electric field lines are highly concentrated in this dielectric layer between the strip and the top of the metallic vias/pads contributing significantly to the most part of the capacitance of the S-MS line. Thus, the capacitance associated with the electric field passing through the air (the air fringing field) can be disregarded. The significant electric field lines to be considered can be separated in regions, called Bottom Plate and Point Charge Fields. The few electric field lines that penetrate between the metallic vias are referred to as penetrating field.

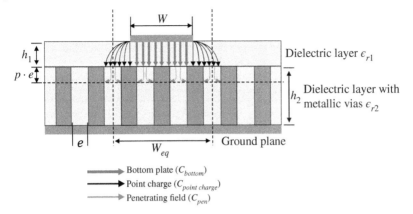

Bottom plate (C_{bottom})
Point charge ($C_{point\ charge}$)
Penetrating field (C_{pen})

Figure 3.9 Representation of electric field lines for capacitance calculation in S-MS lines. The electric field lines associated with the primary component of the linear capacitance are considered in an equivalent width W_{eq}.

Each region of electrical field lines can be attributed a capacitance: C_{Bottom} for Bottom Plate, $C_{Point\ Charge}$ for Point Charge and C_{Pen} for penetrating field, as depicted in Fig. 3.9. The combined capacitance of C_{Bottom} and twice $C_{Point\ Charge}$, arranged in parallel, is denoted as C:

$$C = C_{Bottom} + 2 \cdot C_{Point\ Charge} \tag{3.8}$$

As presented in the modeling method proposed in Zhao et al. (2009), the calculation of C_{Bottom} can be simplified as a parallel plate capacitance from the signal strip to the metallic vias (or via pads) top with separation h_1 and width W.

For the calculation of the capacitance $C_{Point\ Charge}$ in equation (3.9), one must account for an elliptical path for the electrical field lines between the bottom corner of the strip and the metallic vias (or via pads) top, as illustrated in Fig. 3.10.

The expression of $C_{Point\ Charge}$ is given as (Zhao et al., 2009):

$$C_{Point\ Charge}(k) = \varepsilon_{r1}\varepsilon_0 \int \frac{width}{distance} = \varepsilon_{r1}\varepsilon_0 \int_0^{kh_1} \frac{dx}{h_1 \cdot E\left[1 - \left(\frac{x}{h_1}\right)^2\right]} \tag{3.9}$$

In equation (3.9), k is introduced as a ratio, which will be discussed later in this section. The component dx represents the width over which the integral is computed. This width ranges from 0 to kh_1, with h_1 the substrate thickness. The distance from the corner of the signal strip to the width point (ranging from 0 to h_1) is described as following an ellipsoid

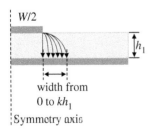

width from
0 to kh_1
Symmetry axis

Figure 3.10 Consideration of electric field lines for calculating the $C_{Point\ Charge}$ capacitance.

path. The term $E[1 - (x/h_1)^2]$ represents the complete elliptic integral of the second kind, which means the ellipse perimeter in one quadrant. Here, the ellipse has a semi-major axis h_1 and a semi-minor axis dx. The expression $1 - (x/h_1)^2$ corresponds to the square of the eccentricity of the ellipse.

Alternatively, C can be extracted in the same way as in a classical microstrip line with a dielectric layer with the same characteristics of D_1 in Fig. 3.2 and a ground plane at the top of the nanowires/vias pads, as presented in Hammerstad and Jensen (1980). In a classical microstrip line, $C_{microstrip}$ can be derived from the calculation of the characteristic impedance $Z_{microstrip}$ and effective relative dielectric constant $\varepsilon_{r_{eff}}$, as:

$$C_{microstrip} = \frac{\sqrt{\varepsilon_{r_{eff}}}}{c_0 \cdot Z_{microstrip}} \tag{3.10}$$

where c_0 is the speed of light in vacuum. The values for $C_{microstrip}$ were extracted from simulations performed in Ansys HFSS.

The results indicate that the capacitance $C_{microstrip}$ is lower than the calculated C from equation (3.8) when using $k = 1$ in equation (3.9) for the capacitance $C_{Point\ Charge}$. This suggests that $C_{Point\ Charge}$ might be overestimated, considering that C_{Bottom} can be easily modeled. Therefore, k should be between 0 and 1 to align C with $C_{microstrip}$ and equation (3.8) needs to be redefined as follows:

$$C(k) = C_{Bottom} + 2 \cdot C_{Point\ Charge}(k) \tag{3.11}$$

The model of the capacitance associated with the penetrating field lines, denoted as C_{Pen}, can use a parallel plate capacitance model with a width W_{eq} indicated in Fig. 3.9. This width, W_{eq}, specifies the region at the top of the nanowires/via pads where the electric field predominates, and is defined as:

$$W_{eq} = W + 2 \cdot k \cdot h_1 \tag{3.12}$$

Then the expression used for C_{pen} is given by:

$$C_{Pen} = \frac{\varepsilon_{r1} \cdot \varepsilon_0 \cdot W_{eq} \cdot (1 - dst_{pad})}{p \cdot e} \tag{3.13}$$

Although the electric field lines that penetrate between the vias do not conform to those of a parallel plate capacitor, the expression for a parallel plate capacitor can still be applied by defining a parallel plate length that accounts for the average length of the field lines. In equation (3.13), $p \cdot e$ accounts for this equivalent length, and it arises from the observation that penetration is proportional to the via pitch e, where p represents a penetration factor derived from measurement results. The density regions where electric field lines penetrate between the vias are accounted by $(1 - dst_{pad})$.

Finally, the linear capacitance of the S-MS line C_{SMS} can be calculated, as shown in equation (3.14), as a sum of two terms. One related to the linear capacitance $C(k)$ in the regions where the electric field lines flow from the strip to the top of the metallic vias/pads, with density dst_{pads}. The other accounts for the penetrating field lines in the regions where the electric field lines flow from the strip to the dielectric between the metallic vias/pads, with density $(1 - dst_{pad})$. This is modeled by two series capacitances, the first related to the electrical field lines in the region of the dielectric layer with ε_{r1} and the second related to the electrical field lines in the region between the vias, C_{pen}.

$$C_{SMS} = dst_{pads} \cdot C(k) + \cfrac{1}{\cfrac{1}{(1-dst_{pads}) \cdot C(k)} + \cfrac{1}{C_{pen}}} \qquad (3.14)$$

This model was validated using fabricated S-MS lines with varying widths in two distinct technologies: PCB and MnM, as detailed further in Section 3.5.8. The linear capacitance of the experimental S-MS lines, $C_{SMS-measured}$, can be extracted from measurements using the same formula as: $C_{microstrip}$ at low frequency, where the capacitance approaches a quasi-static state, thus minimizing dispersion effects caused by the periodicity of the vias:

$$C_{SMS-measured} = \frac{\sqrt{\varepsilon_{r_{eff}}}}{c_0 \cdot Z_{measured}} = \frac{\beta}{Re\{Z\} \cdot 2\pi f} \qquad (3.15)$$

where f represents the designated low frequency, β the phase constant $\left(\beta = Im\left\{\frac{acoshA}{line\ length}\right\}\right)$, Z the characteristic impedance of the S-MS line, calculated as $Z = \sqrt{B/C}$, and A, B, C are the elements of the ABCD matrix derived from measured S-parameters.

A comparison between C_{SMS} calculated from equation (3.14), $C_{microstrip}$ calculated from equation (3.10), and $C_{SMS-measured}$ is provided in Fig. 3.11 within a range of strip widths. The penetration factor p is adjusted for each technology to match $C_{SMS-measured}$. In PCB technology, the penetration factor p was fixed to 0.1. However, in MnM technology, the substantial density of nanowires per area (high dst_{pad}) minimizes significant penetration of the field, rendering C_{Pen} negligible. Consequently, the capacitance is equivalent to that of a classical microstrip line with a substrate height equal to h_1.

From Fig. 3.11, it becomes evident that in PCB technology, the linear capacitance C_{SMS} deviates from $C_{microstrip}$ as it considers the penetrating electric field between pads and vias. Such disparity significantly impacts the characteristic impedance and phase constant β of the S-MS lines, thus justifying this capacitance modeling.

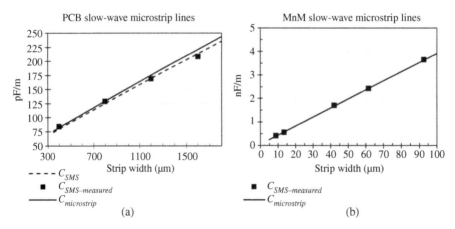

Figure 3.11 Comparison between $C_{SMS-measured}$, $C_{microstrip}$ and C_{SMS} in (a) PCB technology and (b) MnM technology.

3.5.2 Linear Inductance L_{SMS}

As described earlier in this chapter, the magnetic field in S-MS lines is disturbed by the presence of metallic vias. Consequently, the model for the linear inductance L_{SMS} of the S-MS line takes these interactions into account, differing from the linear inductance of classical microstrip lines. In PCB technology, the magnetic field lines pass around the metallic vias as their diameter exceeds the skin depth at microwave frequencies, significantly affecting L_{SMS}. Conversely, in MnM technology, the dimensions of the nanowires are on the nanometer scale, much smaller than the skin depth at millimeter waves; thus, the effect on L_{SMS} is relatively minor.

3.5.2.1 PCB Technology

The magnetic field behavior in a classical microstrip line is illustrated in Fig. 3.12, showcasing two scenarios: in (a), with a dielectric layer thickness of h_1 and in (b), with a thickness of $h_1 + h_2$, where L_1 and L_2 represent their respective linear inductances. The inductance related to the magnetic field flowing only through the part of the substrate with thickness h_2, illustrated in Fig. 3.12 (c), is calculated as the difference $L_2 - L_1$. This portion of the field is equivalent to the portion of the magnetic field flowing through the substrate with vias, in Fig. 3.12 (d), multiplied by a weighting term $(1 - dst_{vias} - \eta)$ that excludes the vias density. η represents a shape factor accounting for the magnetic field line deviation around the metallic vias. The total linear inductance of the S-MS line L_{SMS} is presented in equation (3.16), calculated by summing this term with L_1.

$$L_{SMS} = (L_2 - L_1)(1 - dst_{vias} - \eta) + L_1 \qquad (3.16)$$

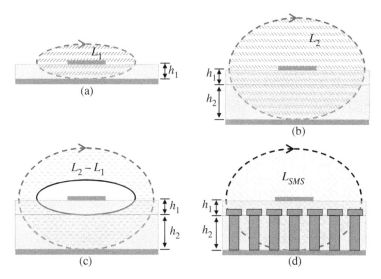

Figure 3.12 Illustration of magnetic field distribution and corresponding inductance for classical microstrips (a) with dielectric thickness h_1, (b) with dielectric thickness $h_1 + h_2$. The flux difference between configurations (a) and (b) is shown in (c). (d) Magnetic field distribution of a S-MS line. Source: Adapted from Hammerstad & Jensen, 1980.

The linear inductance can also be extracted from measurements of the fabricated PCB S-MS lines. The inductance $L_{SMS-measured}$ is extracted at low frequency to mitigate potential dispersion issues that could come from the inductive behavior of the vias, and is given by:

$$L_{SMS-measured} = Re\{Z\}\frac{\beta}{2\pi f} \tag{3.17}$$

The linear inductance of the classical microstrip line illustrated in Fig. 3.12(b), $L_{microstrip}$, can also be calculated from equation (3.17) extracted from simulations performed in Ansys Electronics software using the ABCD parameters in a similar way used for the calculation of the linear capacitance C_{SMS}.

The proposed model is validated through a comparison between L_{SMS}, $L_{SMS-measured}$, and $L_{microstrip}$, shown in Fig. 3.13.

Figure 3.13 demonstrates a strong agreement between the developed expression for L_{SMS} and $L_{SMS-measured}$ across a broad range of signal strip widths. L_{SMS} was fitted using a constant shape factor η of 0.080 for all S-MS lines, regardless of the strip width or the characteristic impedance of those lines. $L_{microstrip}$. Differently, $L_{microstrip}$ overestimates L_{SMS} and $L_{SMS-measured}$ in all cases, showing the convenience of the modeling.

3.5.2.2 MnM Technology

In MnM technology, as the nanowires diameter is smaller than the skin depth at millimeter wave frequencies, the linear inductance L_{SMS} of the S-MS lines in equation (3.16) can be simplified. The term dst_{vias} can be neglected, resulting in the expression given in equation (3.18):

$$L_{SMS} = (L_2 - L_1)(1 - \eta) + L_1 \tag{3.18}$$

The same comparison applied to the lines in PCB technology (Fig. 3.13) is used for the S-MS lines fabricated in MnM technology. Although the influence of the nanowires on the magnetic field is minor, the factor η remains necessary. An empirical factor η equal to 0.067 is used for these S-MS lines, and the results are shown in Fig. 3.14. The same behavior is observed, L_{SMS} agrees very well with the experimental data and $L_{microstrip}$ overestimates both. Naturally, this overestimation is much slighter than the case of PCB technology, given the reduced disturbance to the magnetic field. Nonetheless, the use of the model still yields

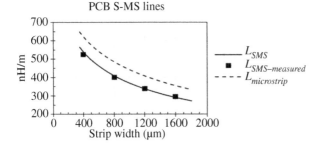

Figure 3.13 PCB technology: comparison between model – L_{SMS} (solid line), measurement – $L_{SMS-measured}$ (squares) of the fabricated S_MS lines, and $L_{microstrip}$ (dotted line) of a classical microstrip line in Fig. 3.12(b).

Figure 3.14 MnM technology: comparison between model – L_{SMS} (solid line), measurement – $L_{SMS-measured}$ (squares) of the fabricated S_MS lines, and $L_{microstrip}$ of a classical microstrip line in Fig. 3.12(b). Source: Adapted from Hammerstad & Jensen, 1980.

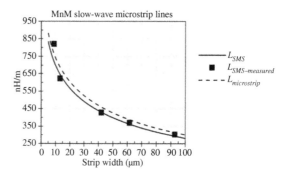

a more accurate estimation of the linear inductance, applicable across a wide range of impedances, since keeping the thickness of the SiO$_2$ layer much smaller than the membrane's thickness.

3.5.3 Linear Strip Resistance R

The linear resistance R modeling the conductor loss od the S-MS line can be calculated from Rautio (2000) as:

$$R\,[\Omega/m] = (1+j)\frac{R_{RF}\,\sqrt{f}}{1 - e^{-(1+j)\frac{R_{RF}\,\sqrt{f}}{R_{DC}}}} \tag{3.19}$$

with $R_{RF} = 0.5\sqrt{\pi\mu_0}/(W\sqrt{\sigma})$ and $R_{DC} = 1/(W\cdot t\cdot\sigma_{copper})$, W the strip width, t the strip thickness, σ the metal conductivity, f the frequency and μ_0 the permeability in free space.

3.5.4 Linear Conductance G

The linear conductance G can be simply calculated from equation (3.20), considering only the upper dielectric layer of the S-MS line in the loss tangent, as the electric field is confined in this layer.

$$G\,[S/m] = \omega \cdot C_{SW} \cdot \tan\delta \tag{3.20}$$

3.5.5 Metallic via Inductance L_{via} and Mutual M_{ij}

The effect of the metallic via itself is significant on the model of the S-MS lines. Additionally, the interaction between them is equally significant. Here, we derive expressions for the inductance L_{via} attributed to the metallic vias or nanowires under the signal strip for a given section with length Δz, as well as the mutual inductance M_{ij} between sections. The value of Δz should be selected considerably smaller than the minimum wavelength used to ensure a steady state can be assumed within each elementary section.

One elementary section of the model is considered a rectangular region along the strip with size $W \cdot \Delta z$, as shown in Fig. 3.15. Within this region, with N nanowires arranged in a hexagonal configuration, as depicted in Fig. 3.3(c), uniform current distribution in the nanowires is assumed.

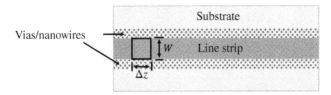

Figure 3.15 Top view of a S-MS line with the region $W \cdot \Delta z$ for the calculi of M_{ij}.

L_{via} can be calculated as a sum form as proposed in Paul (2010) from equation (3.21), with k and l the vias/nanowires within $W \cdot \Delta z$.

$$L_{via}[H] = \frac{\sum_{k=1}^{N} \sum_{l=1}^{N} L_{M_{kl}}}{N^2} \tag{3.21}$$

$$L_{Mkl}[H] = \begin{cases} \frac{\mu_0}{2\pi} h_2 \left[\ln\left(\frac{h_2}{r} + \sqrt{\left(\frac{h_2}{r}\right)^2 + 1} \right) - \sqrt{\left(\frac{r}{h_2}\right)^2 + 1} + \frac{r}{h_2} + \frac{1}{4} \right], & \text{if } k = l \\[3mm] \frac{\mu_0}{2\pi} h_2 \left[\ln\left(\frac{h_2}{d_{kl}} + \sqrt{\left(\frac{h_2}{d_{kl}}\right)^2 + 1} \right) - \sqrt{\left(\frac{d_{kl}}{h_2}\right)^2 + 1} + \frac{d_{kl}}{h_2} \right], & \text{if } k \neq l \end{cases} \tag{3.22}$$

where d_{kl} is the center-to-center distance between nanowire k and nanowire l, and h_2 is the height of the nanowire/via.

In a similar manner, the mutual inductance M_{ij} between two different sections i and j, each one with N nanowires, is calculated as (Paul, 2010):

$$M_{ij}[H] = \frac{\sum_{m=1}^{N} \sum_{n=1}^{N} M'_{mn}}{N^2} \tag{3.23}$$

$$M'_{mn}[H] = \frac{\mu_0}{2\pi} h_2 \left[\ln\left(\frac{h_2}{d_{mn}} + \sqrt{\left(\frac{h_2}{d_{mn}}\right)^2 + 1} \right) - \sqrt{\left(\frac{d_{mn}}{h_2}\right)^2 + 1} + \frac{d_{mn}}{h_2} \right] \tag{3.24}$$

where m a via/nanowire in section i; n a via/nanowire in section j; and d_{mn} the center-to-center distance between nanowire m and nanowire n. Fig. 3.16 shows the significant impact of M_{ij} on modeling the dispersion of the effective dielectric constant ($\varepsilon_{r_{eff}}$) of S-MS line measurements.

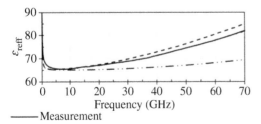

Figure 3.16 Comparison of the effective dielectric constant extracted from measurements (solid line) and from the model with (dashed line) and without (dash/dot line) the mutual coupling between the vias/nanowires.

——— Measurement

- - - - Model with mutual coupling between L_{via}

— ·· — Model without mutual coupling between L_{via}

3.5.6 Metallic vias Resistance R_{via}

The linear via resistance R_{via} can be determined by accounting for the quantity of vias/nanowires of height h_2 over a region defined by $W_{eff} \cdot \Delta z$, from equation (3.25). In this equation, W_{eff} is the equivalent width defined for C_{SMS} in Subsection 3.5.1, and Δz is the model section length. It considers an equivalent conductance, σ, proportional to the density of the vias/nanowires, where $\sigma = dst_{pads} \cdot \sigma_{copper}$.

$$R_{via} [\Omega] = \frac{h_2}{\Delta z \cdot W_{eff} \cdot dst_{pads} \cdot \sigma_{copper}} \tag{3.25}$$

3.5.7 Electrical Model for Coupled Lines

The same parameters can be derived to develop the electrical model of coupled S-MS lines (illustrated in Fig. 3.4) shown in Fig. 3.17. This figure describes the equivalent circuit of a symmetrical coupled S-MS line. As in the model for the S-MS lines, a series resistance and inductance (R and L_{SMS}) model the metallic strips, while C_{SMS} represents the capacitance between the strip and the ground plane (top of the vias/pads). Additionally, mutual inductance (L_M) and mutual capacitance (C_{SS}) are used to represent the magnetic and capacitive coupling between the two strips. For the even-mode (Fig. 3.17(b)), the symmetry plane acts as a magnetic wall (open circuit), and no current flows between the two strips. Note that C_{SS} has no influence on this mode since this capacitance is connected to the same potentials. For the odd-mode (Fig. 3.17(a)), the symmetry plane acts as an electric wall (short circuit) with zero voltage.

C_{SMS} and L_{via} are the capacitance between a signal strip and the upper part of the via pads and the inductive effect induced by the current flowing in the vias array. C_{ss} is the direct coupling capacitance between the two parallel strips. L_{SMS} and L_m are the self-inductance of one strip and the mutual inductance between the two parallel strips, respectively.

As classical coupled lines, the structure of coupled S-MS lines supports two distinct modes of propagation: even-mode and odd-mode. The even- and odd-mode characteristic impedances (Z^e and Z^o) and effective dielectric constants (ε^e and ε^o) of the coupled S-MS

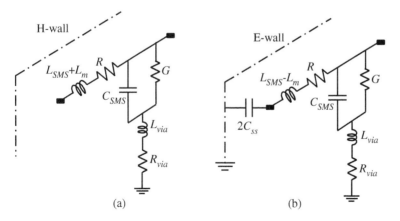

Figure 3.17 Coupled S-MS electrical models of the (a) even-mode and (b) odd-mode.

lines are expressed by equations (3.28) and (3.29), respectively. The electrical (k_c) and magnetic (k_L) coupling coefficients are expressed by equations (3.30).

$$C_e = C'_s, C_o = C'_s + 2C_{ss} \tag{3.26}$$

$$L_e = L_{SMS} + L_m, L_o = L_{SMS} - L_m \tag{3.27}$$

$$Z^e = \sqrt{\frac{L_{SMS} + L_m}{C'_s}}, Z^o = \sqrt{\frac{L_{SMS} - L_m}{C'_s + 2C_{ss}}} \tag{3.28}$$

$$\varepsilon^e = \frac{(L_{SMS} + L_m)C'_s}{c_0^2}, \varepsilon^o = \frac{(L_{SMS} - L_m)\left(C'_s + 2C_{ss}\right)}{c_0^2} \tag{3.29}$$

$$k_c = \frac{C_e - C_o}{C_e + C_o} = \frac{C_{ss}}{C_{ss} + C'_s}, k_L = \frac{L_e - L_o}{L_e + L_o} = \frac{L_m}{L_o} \tag{3.30}$$

where C_e and C_o are the even- and odd-mode linear capacitances, L_e and L_o are the even- and odd-mode linear inductances, c_0 is the speed of light in vacuum, and C'_s corresponds to the equivalent capacitance of C_{SMS} in series with L_{via}. When considering frequencies in the low microwave spectrum, L_{via} does not affect significantly the value of C_{SMS} so that $C'_s \simeq C_{SMS}$. From equation (3.29), the difference between ε^e and ε^o can be derived as:

$$\varepsilon^e - \varepsilon^o = \frac{(L_{SMS} + L_m)C_{SMS}}{c_0^2} - \frac{(L_{SMS} - L_m)(C_{SMS} + 2C_{ss})}{c_0^2} \tag{3.31}$$

and can be simplified as:

$$\varepsilon^e - \varepsilon^o = \frac{2[L_m C_{SMS} - (L_{SMS} - L_m)C_{ss}]}{c_0^2} \tag{3.32}$$

3D electromagnetic simulations can be performed in order to estimate the even- and odd-mode electrical parameters. The two configurations of Fig. 3.17 should be considered, i.e. with perfect H or perfect E boundary conditions. The charts in Fig. 3.18 were derived from these 3D electromagnetic simulations. They give the electrical parameters of several coupled S-MS lines in terms of even- and odd-mode characteristic impedances and effective dielectric constants.

These graphs form the design basis for directional couplers. From the graph in Fig. 3.18 (b) representing the effective relative dielectric constants of even- and odd-modes, it is observed that obtaining close values for even- and odd-modes is not possible. Thus, it is not possible to achieve backward-type couplers, as the isolation would be poor (Bahl et al., 2007). However, by examining the characteristic impedances of the even- and odd-modes, it is noted that it is possible to obtain very close values around 50 Ω by using a wide gap G. Thus, coupled S-MS lines can lead to the realization of forward-type directional couplers with a very interesting length due to the significant difference between ε^e and ε^o. The design of forward-type directional couplers is the subject of Section 3.6.3 in this chapter.

3.5.8 Validation

To verify the electrical model described earlier, the topology shown in Fig. 3.8 was used as an elementary segment of the distributed model of a S-MS line. Multiple segments were cascaded to simulate the entire transmission line in ADS-Keysight software until the

Figure 3.18 Even- and odd-mode electrical parameters versus strips widths *W* and spacing *G*, at 4.5 GHz. Black dots: EM simulation points (a) Characteristic impedances, (b) Effective dielectric constants.

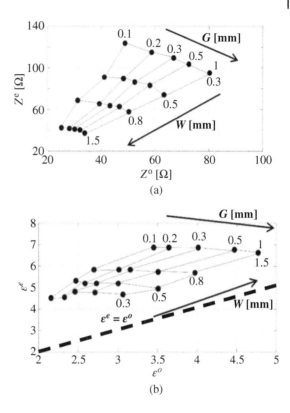

(a)

(b)

S-parameters within the selected frequency range exhibited minimal changes. The values of each component within the segment were determined from the equations provided in 3.5.1–3.5.6.

Afterward, the characteristic impedance Z_0, attenuation constant α, and effective dielectric constant $\varepsilon_{r_{eff}}$ of S-MS lines were derived from these simulations. These values were then compared with those obtained from measurements of fabricated S-MS lines in PCB and MnM technologies, as well as 3D electromagnetic simulations performed in Ansys HFSS. The results are presented here separately for each technology.

3.5.8.1 PCB Technology

An example of a fabricated PCB S-MS line is given in Fig. 3.19. The substrate is composed of a top and bottom copper layer and three dielectric layers in between: RO4003 in the bottom and top layers, and RO4450 in the intermediate layer illustrated in Fig. 3.19(a). Figure 3.19(b) shows the fabricated S-MS line, where the signal strip has a length of 10 mm and width of 400 µm. Figure 3.19(c) shows the bottom of the S-MS lines with the vias (with pads) arranged in five rows along the line. To achieve a hexagonal configuration shown in Fig. 3.3, two rows were offset. The center axis of the rows aligns with the center of the signal strip. For measurement purposes, this S-MS lines are fed by grounded coplanar waveguide (GCPW). Copper is used as metal with thickness of 35 µm. The vias have a pitch $e = 400$ µm and a diameter $d = 400$ µm, while the pads have a pitch $e_{pad} = 200$ µm and

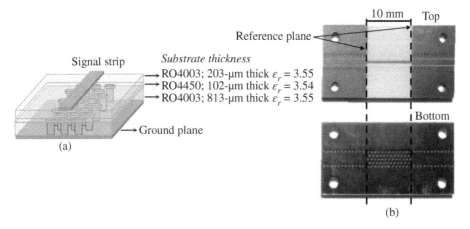

Figure 3.19 Fabricated S-MS line in PCB technology. (a) Illustration of the substrate used, (b) top and bottom views of the S-MS line with grounded CPW feeding lines.

diameter $d_{pad} = 600\,\mu m$, resulting in a vias area density $dst_{vias} = 22.7\%$ and pads area density $dst_{pads} = 51.1\%$.

The PCB S-MS lines were measured in an Anritsu 3739D vector network analyzer (VNA). Calibration was performed using a thru-reflect-line (TRL) method, with reference planes positioned after the feeding lines, as shown in Fig. 3.19(b). As $L_{SMS-measured}$ is extracted at low frequencies, the frequency effect was included in Z_0 through (Marks & Williams, 1991):

$$Z_0 = \frac{R + j\omega \cdot L_{SMS-measured}}{\gamma_{measured}} \qquad (3.33)$$

where $\gamma_{measured}$ is the propagation constant extracted from measured S-parameters.

For the calculation of the model parameters, first the factors p, η and Δz must be determined. A single simulation for a particular signal strip width suffices to extract the penetration factor p (for the linear capacitance C_{SMS} in Section 3.5.1) and the shape factor η (for the linear inductance L_{SMS} in Sections 3.5.2), as the two factors remain constant across varying signal strip widths. The factors can also be extracted from measurements, and p was set to 0.1 and η was set to 0.08. For Δz, as the maximum operating frequency of the PCB S-MS lines is approximately 10 GHz, before significant dispersion occurs due to the inductive behavior of the vias, we can calculate $\lambda_g = 12.5$ mm. This value considers a maximum $\varepsilon_{r_{eff}}$ of 5.8 – from Fig. 3.20 at DC. Following the classical rule of thumb, $\frac{\Delta z}{\lambda_g}$ should be less than 1/20. Thus, Δz must be smaller than 625 µm. However, considering that a "column" with five vias is the vias periodicity and equals 833 µm, to ensure that at least one column of vias is included in each section, Δz is set to 833 µm. This leads to $\frac{\Delta z}{\lambda_g} = 1/15$, which is still acceptable. For instance, a S-MS line with 12 sections, each being 833-µm long, has a length of 10 mm.

Figure 3.20 shows the parameters extracted from measurements, model, and simulations performed in Ansys HFSS. The access lines were not considered in the simulations, since they were compared with de-embedded measurements.

Figure 3.20 Comparison between the characteristic impedance Z_0, attenuation constant α, relative effective dielectric constant $\varepsilon_{r_{eff}}$ extracted from measurements, model, and simulation for S-MS lines with different widths in PCB technology.

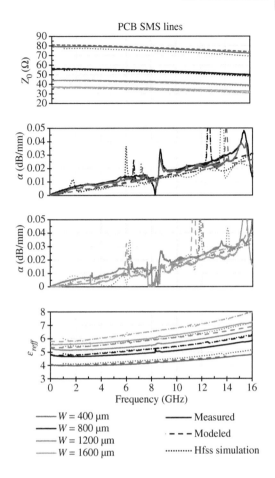

The general agreement of the model with the measurements and simulations is highly satisfactory, indicating the strong accuracy of the electrical model for S-MS lines in PCB technologies.

3.5.8.2 MnM Technology

Here, an example of a S-MS line fabricated in MnM technology is given in Fig. 3.21. The layer of the signal strip and the ground plane are made of copper (with 1.5 μm of thickness), as well as the nanowires. The dielectric around the nanowires is alumina and the dielectric layer between the alumina and the strip signal is made of SiO_2.

Figure 3.21(b) shows the fabricated S-MS line featuring a signal strip measuring 500 μm in length and 92.75 μm in width. The figure also shows the ground-signal-ground (GSG) pads for measurements. The ground pads are interconnected and connected to the ground plane through nanowires. The substrate employed is an AAO membrane from InRedox, with nanowire diameter $d = 40$ nm and pitch $e = 67$ nm.

The simulation of the MnM S-MS lines requires a very high computational cost, considering the numerous nanowires. To mitigate this cost, larger nanowires and spacing were used in simulations, preserving the original via area density dst_{via} of 12%.

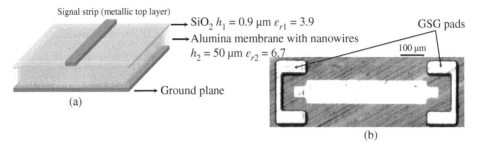

Signal strip (metallic top layer)
SiO_2 $h_1 = 0.9$ μm $\varepsilon_{r1} = 3.9$
Alumina membrane with nanowires
$h_2 = 50$ μm $\varepsilon_{r2} = 6.7$
GSG pads
100 μm
Ground plane
(a)
(b)

Figure 3.21 Fabricated S-MS line in MnM technology. (a) Illustration of the substrate used, (b) top view of a S-MS line along with ground-signal-ground (GSG) measurement pads (with interconnected grounds).

The Keysight N5227B VNA was used to measure the MnM S-MS lines through MPI Titan probes with 100-μm-pitch. The calibration used a load-reflect-reflect-line (LRRM) method, and the lines were de-embedded using the two-line method described in Mangan et al. (2006).

The same procedure used for PCB S-MS lines in the previous section was used in this MnM case. The characteristic impedance Z_0 was calculated from the same equation (3.3) and a shape factor η of 0.067 was used to calculate the linear inductance L_{SMS}. However, in the MnM technology the penetration of the electrical field between the nanowires is negligible; therefore no p factor is necessary. The maximum operating frequency is 100 GHz, and the highest $\varepsilon_{r_{eff}}$ is 98 – from Fig. 3.22 at DC. Thus, $\frac{\Delta z}{\lambda_g}$ less than 1/20 is satisfied by $\Delta z = 15$ μm, which means a S-MS lines with 100 sections in a length of 1.5 mm. Although the measurements were limited to 70 GHz, a higher frequency of 100 GHz was considered, as these S-MS lines are capable of operating at elevated frequencies.

Figure 3.22 shows the parameters extracted from measurements, models, and Ansys HFSS simulations.

The agreement of the characteristic impedance and attenuation constant α of the MnM S-MS lines model with measurements and simulations is very accurate. The relative effective dielectric constant $\varepsilon_{r_{eff}}$, agrees very well up to around 30 GHz, but above that, it slightly overestimates the dispersion. Although the simulations can predict Z_0 and $\varepsilon_{r_{eff}}$ effectively, the attenuation constant is greatly overestimated due to the presence of eddy currents flowing in the larger nanowires and larger pitch used in the simulations.

Analysis of the Oxide Thickness Here, several other S-MS lines that were fabricated and characterized are presented. They are based on MnM-substrates with two different SiO_2 dielectric layer thicknesses (t_{ox}): one with 0.8 μm (t_{ox1}) and the other, with 1 μm (t_{ox2}). The membrane thickness (H) of both substrates is 50 μm with 55-nm pore diameter and 143 nm of interpore distance. The metal thickness of the strips, ground pads, and ground plane is 3 μm. Figure 3.23 shows a fabricated sample with several groups of S-MS lines having different lengths (d) and same signal strip width (W). Each group, shown in the inset of Fig. 3.23, differs in signal strip width.

The ground pads needed for measurement are capacitively coupled to the ground plane through the dielectric SiO_2 layer, hence avoiding the need for vias.

Figure 3.22 Comparison between the characteristic impedance Z_0, attenuation constant α, relative effective dielectric constant $\varepsilon_{r_{eff}}$ extracted from measurements, model, and simulation for S-MS lines with different widths in MnM technology.

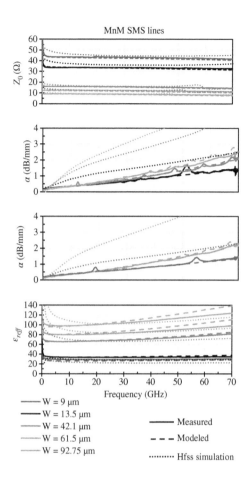

— W = 9 μm
— W = 13.5 μm
— W = 42.1 μm
— W = 61.5 μm
— W = 92.75 μm

— Measured
– – Modeled
·········· Hfss simulation

Figure 3.23 S-MS lines fabricated on the MnM-substrate. Source: Ariana L. C. Serrano, 2014/with permission of IEEE.

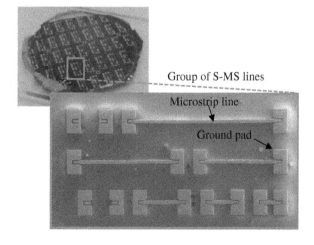

Group of S-MS lines

Microstrip line

Ground pad

The electrical parameters of the S-MS line, such as characteristic impedance (Z_c), effective relative dielectric constant ($\varepsilon_{r_{eff}}$), and attenuation constant (α) were extracted from the de-embedded measurement data (S-parameters) (Mangan et al., 2006). Their physical dimensions W and d were measured on a scanning electron microscope. The measured parameters in frequency for two pairs of S-MS lines having similar signal strip widths (W) on the MnM substrate with the two different oxide thickness t_{ox1} and t_{ox1} are given in Fig. 3.24. The summary of these measurements is given in Tables 3.1 and 3.2 at 60 GHz for each width and oxide thickness.

These results show that the thinnest the microstrip width, the higher the characteristic impedance. Thus, higher Z_c can be realized by reducing W, which can be approximately

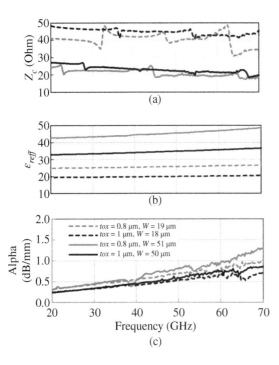

Figure 3.24 Measurements of the characteristic impedance (a), effective relative dielectric constant (b), and attenuation constant (c), versus frequency, for two W values and two t_{ox} values.

Table 3.1 Summary of measurements of the slow-wave transmission lines at 60 GHz with $t_{ox} = 0.8$ μm.

W (μm)	Z_c (Ω)	$\varepsilon_{r_{eff}}$	α (dB/mm)	Q
14	40	22	0.70	36
15	40	24	0.80	32
19	43	26	0.80	34
20	36	28	0.83	35
36	23	39	1.00	36
45	20	44	0.93	37
51	19	47	0.95	39

Table 3.2 Summary of measurements of the slow-wave transmission lines at 60 GHz with $t_{ox} = 1.0$ μm.

W (μm)	Z_c (Ω)	$\varepsilon_{r_{eff}}$	α (dB/mm)	Q
18	43	20	0.62	37
21	38	23	0.65	38
28	31	26	0.70	39
34	28	30	0.71	40
38	25	31	0.72	40
41	24	32	0.73	43
45	23	35	0.74	44
50	22	36	0.75	47

$100\,\Omega$ for a 2 μm-width signal strip. This can also be realized by processing specific regions without the metallic nanowires in the membrane. Without the nanowires, the distance between the strip and the ground plane increases (from the electric field point of view), decreasing the capacitance and thus increasing Z_c.

It is possible to see in Fig. 3.24 that the parameters slightly change with frequency: $\varepsilon_{r_{eff}}$ and α increase with frequency, whereas Z_c decreases. α increases exponentially. These effects are directly related to the metallic nanowires, as confirmed by the model of the S-MS lines. The losses of the S-MS lines are between 0.6 and 1 dB/mm at 60 GHz. This value, which is obtained on a low-cost substrate, is very comparable to the insertion loss in conventional microstrip lines on well-controlled integrated technologies like CMOS (Cathelin et al., 2007).

The quality factor $Q = \beta/(2\alpha)$ is higher than 32 and up to 47 at 60 GHz. When the capacitance is maximized, i.e., when the signal strip width is large and the oxide thickness is thin, $\varepsilon_{r_{eff}}$ reaches more than 40. This highlights the high miniaturization potential of S-MS lines on the MnM substrate: if the substrate was entirely alumina ($\varepsilon_r = 9.8$) without the nanowires or nanopores, and considering the relative thin SiO_2 layer negligible, conventional microstrip lines would present $\varepsilon_{r_{eff}} \sim 5.7$ for $W = 18$ or 19 μm and $\varepsilon_{r_{eff}} \sim 6.3$ for $W = 50$ or 51 μm.

The oxide thickness variation changes the capacitance of the S-MS line, barely changing its inductance. Then, the characteristic impedance and the effective relative dielectric constant can be adjusted by choosing the appropriate oxide thickness and signal strip width. It is interesting to note that for the two different oxide thicknesses t_{ox1} and t_{ox2}, the associated electrical parameters (Z_{c_tox1}, Z_{c_tox2}, $\varepsilon_{r_{eff}\_tox1}$, $\varepsilon_{r_{eff}\_tox2}$) are related to the oxide thickness ratio by equation (3.34) and (3.35). This is experimentally confirmed in Fig. 3.25, where the calculus of Z_{c_tox2} and $\varepsilon_{r_{eff}\_tox2}$ is also plotted. The agreement between this calculation and the measured characteristic impedance for a 1 μm-thick oxide is very good on the whole signal strip width range.

$$\frac{Z_{c\_tox1}}{Z_{c\_tox2}} = \sqrt{\frac{t_{ox1}}{t_{ox2}}} \tag{3.34}$$

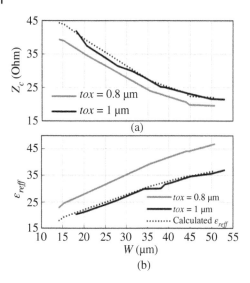

Figure 3.25 Characteristic impedance (a) and effective relative dielectric constant (b) versus the signal strip width for two different oxide thicknesses. The calculated values for Z_c and $\varepsilon_{r_{eff}}$ were obtained by equations (3.34) and (3.35).

$$\frac{\varepsilon_{r_{eff}-tox1}}{\varepsilon_{r_{eff}-tox2}} = \frac{t_{ox2}}{t_{ox1}} \tag{3.35}$$

The S-MS lines were simulated in ADS, from Keysight, cascading several sections of the model with the values extracted from the model. Figure 3.26 shows the simulation and measurement results for two signal strip widths on the same substrate.

The simulations used different values for all elements of the circuits, since the S-MS lines are different in width. A larger signal strip width increases the linear capacitance C, reduces R, and reduces both R_{via} and L_{via}, as they are electrically in parallel.

While the series resistance decreases, the attenuation constant increases with larger signal strip widths, because Z_c decreases as the linear inductance L decreases and C increases (as for a conventional microstrip line). This can be seen in Figs. 3.25(a), 3.26(a), and 3.27(a), and is simply explained by equation (3.36), where the first term and second term are the metallic and dielectric losses, respectively. Because both alumina and SiO_2 are known for having very small dielectric loss tangents, the second term considering G, the parallel conductance in a low-loss $RLGC$ model, is negligible.

$$\alpha = {}^R\!/_{Z_c} + GZ_c \tag{3.36}$$

Typically, the dielectric loss tangent of the alumina is in the order of 10^{-4} (Blair et al., 1994) and about 10^{-3} for the SiO_2 (Li et al., 2013), which can vary significantly depending on the deposition techniques. Figure 3.27 shows the simulation and measurement results for two S-MS lines with similar width and different oxide thicknesses.

In this case, almost only C has changed, decreasing with the increase of t_{ox}. L slightly increases, as the inductance increases with the increase in the total thickness of the substrate (t_{ox} + membrane thickness). The attenuation constant is increased with the reduction of t_{ox} since Z_c decreases.

Figure 3.26 Simulation results (dashed line) from the equivalent electrical circuit and measurements (full-line) of the characteristic impedance (a), relative effective dielectric constant (b), quality factor and attenuation (c), of the S-MS lines for two different widths (14 and 45 µm) on the same substrate ($t_{ox} = 0.8$ µm).

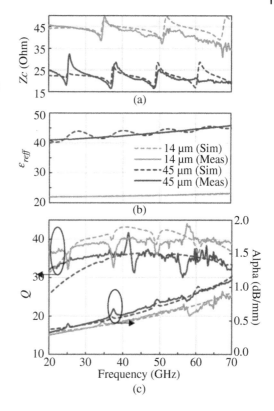

(a)

(b)

- ---- 14 µm (Sim)
- —— 14 µm (Meas)
- ----- 45 µm (Sim)
- —— 45 µm (Meas)

(c)

Air-Filled Suspended Slow-Wave Microstrip Line AF-S-MS The MnM S-MS lines presented so far use a SiO$_2$ layer as dielectric; thus, the dielectric losses are associated with the loss tangent of the SiO$_2$ film. The MnM substrate fabrication process requires low-temperature (<200 °C) SiO$_2$ films, which tend to have higher losses. In order to reduce the attenuation constant, new S-MS lines with suspended signal strips in air instead of using the SiO$_2$ film were developed. Figure 3.28 illustrates the air-filled suspended S-MS (AF-S-MS) in a 3D view. A longitudinal cut-view of the structure is also presented in this figure. Copper nanowires were grown selectively only underneath the suspended strip that is L_s long.

The characteristic impedance Z_0 of the AF-S-MS line is controlled by the width of the suspended strip and its height, indicated by the air gap g. The strip has to be periodically anchored to the substrate to avoid stiction because of static charges or surface tensions of solvents used during the fabrication process. Naturally the anchoring will add some parasitic effects that will interfere with the functioning of the desired suspended strip, so the electrical length of the anchoring has to be kept as small as the fabrication process allows. The final structure, thus, is a periodic series of long, suspended sections with short anchoring sections between them. shows a simplified 3D view of the suspended microstrip line section and a longitudinal cross-section that shows both the suspended regions as well as the anchoring. A mechanical anchor is required for mechanical stability of the structure and for positioning the RF pads used for measurement purposes.

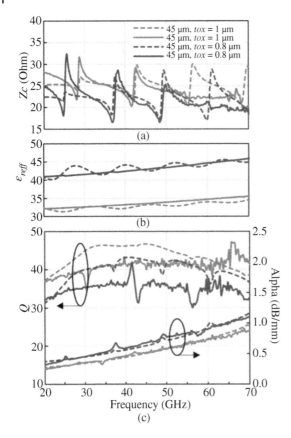

Figure 3.27 Simulation results (dashed line) from the equivalent electrical circuit and measurements (full-line) of the characteristic impedance (a), effective relative dielectric constant (b), quality factor and attenuation constant (c), of the S-MS lines for same widths (45 µm) and different oxide thicknesses (0.8 and 1 µm).

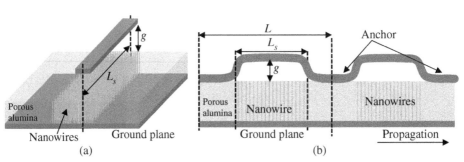

Figure 3.28 Suspended slow-wave transmission line in the MnM substrate. (a) tridimensional view (b) lateral view.

The suspended S-MS lines were designed by 3D EM simulations. Due to the presence of air below the signal strip, only S-MS lines with high Z_0 is possible. Lines with signal strip width of 15, 25 and 35 µm lead to Z_0 of 65, 75 and 90 Ω, respectively.

From the mechanical aspect, two suspended lengths L_s were considered: 250 and 500 µm. Longer lengths are possible if the intrinsic stress of the top copper layer (the signal strip) is small. Longer L_s is desirable, since it reduces the number of the 50 µm long anchors

Figure 3.29 Fabricated suspended
S-MS line with four suspended line
segments and three anchors with its
access pads (a) top view; (b) 3D view.

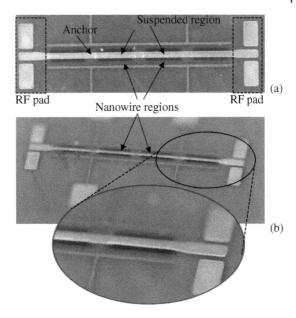

Figure 3.29 Fabricated suspended S-MS line with four suspended line segments and three anchors with its access pads (a) top view; (b) 3D view.

required, thus increasing the slow-wave effect. S-MS lines with one, two and four suspended regions were designed in order to evaluate their performance.

A fabricated suspended S-MS line with four 250-μm-long suspended line segments and three anchors with access pads is shown in Fig. 3.29. Due to the translucid aspect of the AAO, in this figure, it is also possible to see the nanowires (dark areas) underneath the suspended regions.

As expected, no stiction can be seen in the 250 and 500-μm-long segments. No buckling due to possible compressive residual intrinsic stress was present in the fabricated structures. This means that longer segments could indeed be fabricated using the same process.

The suspended S-MS lines were characterized up to 70 GHz using GSG probes (from MPI company, Titan model probes with 100 μm-pitch) in a manual probe station and a Keysight PNA N5227B VNA. A LRRM calibration was used. The frequency responses of three different suspended S-MS lines are presented in Fig. 3.30, showing their insertion and return loss.

Each measured S-MS line is formed by: two suspended line segments of 500 μm with a total length considering the anchors $L = 1150$ μm; four segments of 250 μm ($L = 1250$ μm); and four segments of 500 μm ($L = 2250$ μm).

The structures were designed with capacitive GSG RF pads, i.e. the ground pads were not resistively connected to the ground plane on the bottom of the structure, but through a capacitor. This capacitor is formed between the metal ground pads themselves and the nanowires, separated by the thin SiO_2 layer at 200 nm. With a high capacitance value, these capacitors become a short circuit at low frequencies around 7 GHz, when they electrically connect to the ground plane. Therefore, the measurements are only valid for frequencies above 7 GHz, presenting an unstable behavior below that frequency, as shown in Fig. 3.30. The extracted Z_0 for the three S-MS lines is approximately 65, 75 and 90 Ω, respectively, as designed.

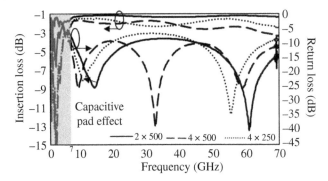

Figure 3.30 Measured frequency response (S-parameters) of three fabricated suspended S-MS lines. The transmission line with two segments of a 500-μm suspended line (solid line); four segments of a 500-μm suspended line (dashed line); and four segments of a 250-μm suspended line (dotted line).

The attenuation constant α, the effective relative dielectric constant $\varepsilon_{r_{eff}}$ and the quality factor ($Q = \beta/[2\alpha]$), where β is the phase constant, were also extracted from the measurements and are shown in Fig. 3.31. As expected, the attenuation constant of all suspended S-MS lines is lower than all the previous ones. It is relevant to indicate that these values are not de-embedded, thus it is expected slightly better results. The slow-wave effect is confirmed, where $\varepsilon_{r_{eff}}$ ranges from 5 to 7.5 compared to a suspended classical microstrip line without slow-wave effect. A stronger slow-wave effect can be obtained by reducing the air gap, using a thinner sacrificial material. The quality factor of these suspended S-MS lines is comparable to or even better than microstrip lines on well-controlled integrated technologies like CMOS (Cathelin et al., 2007).

The suspended S-MS lines can be used to realize MEMS shunt switches and phase shifters, and continuously-tunable phase shifters utilizing reconfigurable substrates, such as liquid crystal, that will be shown in the applications Section 3.6 of this chapter.

3.5.9 Discussion

The electrical model accurately predicts the dispersion of S-MS lines in frequency, particularly evident in the extracted $\varepsilon_{r_{eff}}$. However, the model tends to overestimate dispersion because it does not account for frequency dispersion in L_{SMS}, L_{via}, and M_{ij}. Considering frequency dispersion for these parameters would lead to a decrease in their values with frequency, resulting in an $\varepsilon_{r_{eff}}$ that increases less with frequency, aligning better with measurements. As shown in Fig. 3.16, the mutual coupling represented by M_{ij}, in Fig. 3.8, is crucial for accurately predicting the variation of $\varepsilon_{r_{eff}}$ with frequency. This coupling between sections of the model constitutes the primary feature of the S-MS lines electrical model.

A large range of characteristic impedances can be realized by changing the dielectric thickness of the S-MS lines, without reaching critical dimensions, keeping the fabrication process simple and low-cost. With comparable quality factors and effective dielectric constant, the lateral dimension of S-MS lines is much narrower than their Slow-wave CPW counterparts, leading to a much smaller area. S-MS lines also offer the overall flexibility of classical microstrip lines, in particular, they can be meandered in order to further control

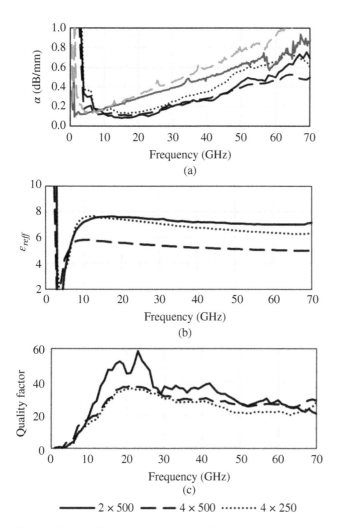

Figure 3.31 (a) Attenuation constant α, (b) effective relative dielectric constant $\varepsilon_{r_{eff}}$ and (c) quality factor of the three fabricated suspended slow-wave transmission lines. The transmission line with two segments of a 500-μm suspended line (solid line); four segments of a 500-μm suspended line (dashed line); and four segments of a 250-μm suspended line (dotted line).

the form factor of passive devices, whereas this is not possible (or is complicated) with slow-wave CPW.

3.6 Applications

In this section, three components performing different functions are presented. The components are designed based on S-MS lines in order to obtain compact designs and good electrical performance. Three different components are presented and characterized: Wilkinson power divider, branch-line coupler, and forward-wave coupled-line directional coupler.

3.6.1 Wilkinson Power Divider

Design examples of 1:2 and 1:4 Wilkinson power dividers (WPDs) operating at 2.45 GHz are shown in this section using S-MS lines in PCB technology. The slow-wave concept created by inserting blind via holes as described in the previous sections describing the PCB technology (3.3 and 3.5.2.1) was applied to the compact WPD topology presented in Burdin et al. (2016). The topology is considered for its simplicity, good performance, and compatibility with the S-MS lines. As described earlier in those PCB sections, the S-MS WPD substrate stack is composed of three dielectric layers. Rogers RO4003 ($\varepsilon_r = 3.55$ and a loss tangent $\tan\delta = 0.0027$) is used in the upper and lower dielectric layers and they are glued with an intermediate prepreg layer RO4450 ($\varepsilon_r = 3.52$ and $\tan\delta = 0.0041$), forming the intermediate layer. The thickness of the upper, intermediate, and lower dielectric layers are 813, 102 and 203 μm, respectively. Via holes are inserted within the lower substrate connected to the bottom ground plane and the pads are placed in the intermediate layer.

Figure 3.32 shows a top view of the proposed designs. Figure 3.32(a) shows the 1:4 divider formed by cascading the 1:2 divider. Figure 3.32(b) shows the 1:2 divider and its parameters (characteristic impedance Z_i and physical length L_i). The three 1:2 dividers used in the layout of the 1:4 divider are identical.

In the layout of the 1:2 divider, the via holes and pads are represented by dashed circles. Due to measurement constraints, a minimum space of 15 mm is considered between the output ports (A and B). Two stubs (stub and mushroom stub) are used to improve the return loss. An SMD resistor (100 Ω) is used to isolate the output ports connected to two short transmission lines of characteristic impedance $Z_3 = 49\,\Omega$.

The values of the 1:2 divider design parameters optimized for an operating central frequency of 2.45 GHz are given in Table 3.3.

Figure 3.33 shows a photograph of the fabricated 1:2 and 1:4 WPDs (S-MS and classical). The area occupied by the 1:2 S-MS WPD (excluding the access lines) is $(9 \times 10)\,\text{mm}^2$. A classical 1:2 WPD based on conventional microstrip lines would occupy an area of $(10 \times 16)\,\text{mm}^2$ on the substrate stack. Hence, the use of the S-MS technique leads to a

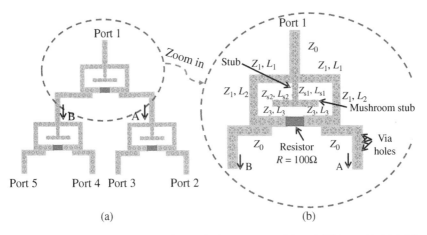

Figure 3.32 Layout of the S-MS (a) 1:4 Wilkinson power divider (WPD) and (b) 1:2 WPD with its parameters.

Table 3.3 Design parameters values of the 1:2 divider: characteristic impedances (Ω) and lengths (mm).

Z_0	Z_1	Z_3	Z_{s1}	Z_{s2}	L_1	L_2	L_3	L_{s1}	L_{s2}
50	65	49	112	88	4.5	7	3.4	3	3.7

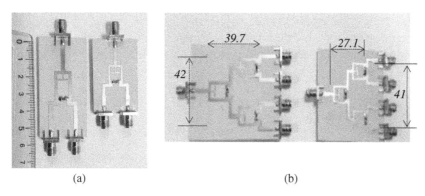

(a)   (b)

Figure 3.33 Fabricated WPDs (a) classical and using S-MS 1:2 divider and (b) classical and using S-MS 1:4 divider. Units in mm.

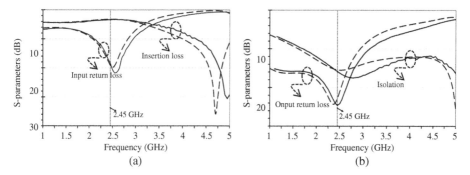

(a)   (b)

Figure 3.34 Simulation (dashed lines) and measurement (solid lines) results of the S-MS 1:2 WPD: (a) input return loss and insertion loss, and (b) output return loss and isolation.

surface miniaturization of 44%. On the other hand, the overall surface area occupied by the 1:4 S-MS WPD is (41×27.1) mm^2 which is 33% less in size than a 1:4 conventional WPD $(42 \times 39.7$ mm$^2)$.

The scattering parameters response of the 1:2 WPD is shown in Fig. 3.34. The response is plotted from 1 to 5 GHz. At the operating frequency of 2.45 GHz, the measured insertion loss is 0.25 dB (above the 3 dB split ratio), the input return loss is better than 19 dB and the output return loss is better than 22 dB. An isolation of 17 dB between the output ports is achieved. On the other hand, the WPD designed using conventional microstrip lines presents an insertion loss of 0.15 dB. The slight increase of insertion loss when the S-MS technique is used, is due to the lower quality factor of the S-MS lines.

Figure 3.35 shows a plot of the S-parameters of the 1:4 S-MS WPD from 2 to 3 GHz. At the operating frequency, the measured insertion loss is 0.3 dB. The input return loss is better than 19 dB and the output return loss is better than 16 dB. The 1:4 WPD based on conventional lines exhibits similar performance at its operating frequency. The measured insertion loss is 0.25 dB, the measured input return loss is better than 18 dB and output return loss is better than 22 dB.

Figure 3.36 shows the isolation between different output ports better than 19 dB and the measured phase and amplitude imbalances of the S-MS 1:4 WPD. The phase imbalance is 1.7° and the amplitude imbalance is 0.21 dB.

A comparison to the state-of-the-art leads to conclude that this method is a good compromise between compactness and electrical performance, offering a new way of simple design for miniaturized circuits.

3.6.2 Branch-Line Coupler

The topology of the proposed S-MS branch-line coupler is presented in Fig. 3.37. The substrate stack combination is composed of a lower dielectric layer using Rogers RO4003 ($\varepsilon_r = 3.55$ and $\tan\delta = 0.0027$) with thickness of 0.813 mm, while the upper dielectric layer is Rogers RO4403 ($\varepsilon_r = 3.4$ and $\tan\delta = 0.005$) with thickness of 0.29 mm. Only one row of metallic vias is embedded in the middle of the lower dielectric layer along the strips length.

S-MS lines with characteristic impedances of 35 and 50 Ω were designed, fabricated and tested. For each S-MS line, three different lengths (16, 32 and 64 mm) were considered in order to extract their electrical characteristics. The width of 2.9 mm is necessary to achieve a 35 Ω characteristic impedance over the prescribed substrate stack. On the other hand, the 50 Ω S-MS transmission line will have a width of 1.4 mm.

In order to validate the values, the lines are measured and the characteristic impedance is extracted using the two-line extraction method (Mangan et al., 2006). Figure 3.38 shows the simulated and measured characteristic impedance and effective relative permittivity $\varepsilon_{r_{eff}}$ for both transmission lines. For a slow-wave transmission line having a characteristic impedance of 35 Ω, the measured value of $\varepsilon_{r_{eff}}$ at 2.45 GHz is 4.3 while the value of $\varepsilon_{r_{eff}}$ of its microstrip counterpart is 2.85. This means that using the via holes layer leads to a SWF equal to $\sqrt{\dfrac{\varepsilon_{r_{eff}-SlowWave}}{\varepsilon_{r_{eff}-Microstrip}}} = \sqrt{\dfrac{4.3}{2.85}} = 1.23$. The measured $\varepsilon_{r_{eff}}$ is 4.2 for the 50 Ω slow-wave transmission line, while the measured $\varepsilon_{r_{eff}}$ of its microstrip counterpart is 2.6, leading to a SWF equal to 1.27.

The 35 and 50 Ω slow-wave transmission lines presented are used for the design and fabrication of the proposed branch-line coupler. The transmission lines physical dimensions are as follows: $W_{35} = 2.9$ mm, $L_{35} = 15.6$ mm, $W_{50} = 1.4$ mm and $L_{50} = 17$ mm.

A photograph of the fabricated S-MS branch-line coupler with a comparison between the simulated and measured magnitude responses of the coupler is shown in Fig. 3.39. The measured values of the insertion loss (S_{12}), coupling coefficient (S_{13}), return loss (S_{11}) and isolation (S_{14}) at 2.36 GHz are 3, 3.1, 25, and 27.5 dB, respectively. The measured bandwidth is 0.6 GHz. The measured phase responses of the inserted and coupled waves are also shown in this figure. The measured phase difference between the output ports is 91° at the

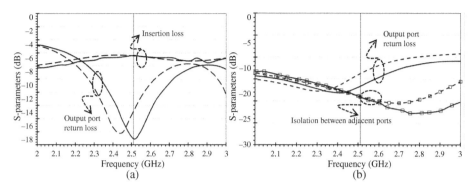

Figure 3.35 Simulation (dashed lines) and measurement (solid lines) results of the S-MS 1:4 WPD.

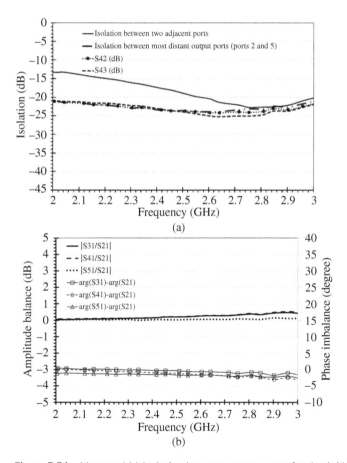

Figure 3.36 Measured (a) Isolation between output ports for the 1:4 WPD and (b) phase and amplitude imbalances.

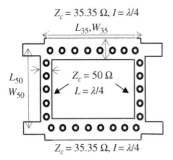

$Z_c = 35.35\ \Omega,\ l = \lambda/4$

L_{35}, W_{35}

L_{50}
W_{50}

$Z_c = 50\ \Omega$
$L = \lambda/4$

$Z_c = 35.35\ \Omega,\ l = \lambda/4$

Figure 3.37 Schematic view of the proposed S-MS branch-line coupler.

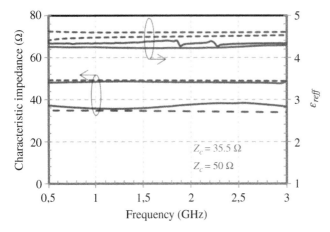

Figure 3.38 Characteristic impedance and relative effective dielectric constant of the S-MS lines (dashed line – Simulated, solid line – Measured).

operating frequency. Hence, the amplitude and phase imbalances remain less than 0.3 dB and 1° dB, respectively over all the operating bandwidth. The surface area of the compact coupler is $17 \times 19.9\ \text{mm}^2$. This surface is 43% less as compared to the coupler realized using classical microstrip lines.

Therefore, a significant reduction in surface area and improvement in electrical performance is noticed when the slow-wave concept is applied to the hybrid branch-line coupler. These results highlight the potential of S-MS lines for the realization of RF compact circuit devices.

3.6.3 Forward-Wave Directional Coupler

A 0-dB forward-wave directional coupler (FWDC) operating at 4.5 GHz was designed using coupled S-MS lines (see Section 3.3.2) in PCB technology. Figure 3.40 shows a photograph of the coupler fabricated in the substrate technology depicted in Fig. 3.3.

The coupled strips width W is 0.8 mm wide, separated by a spacing G of 0.3 mm. The even- and odd-modes characteristic impedances Z^e and Z^o obtained from the chart given in Fig. 3.18(a) are 65 and 43 Ω, respectively, leading to a port matching impedance of $\sqrt{Z^o \cdot Z^o} = 53\ \Omega$. By considering the effective dielectric constants chart of Fig. 3.18(b), ε^e

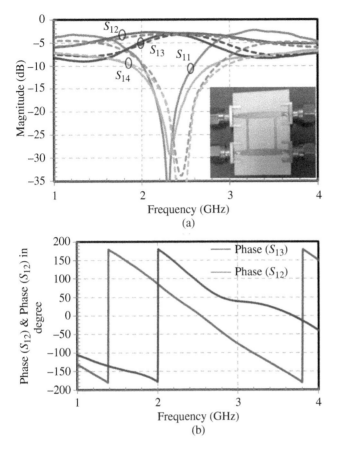

Figure 3.39 S-parameters of the S-MS branch-line coupler: (a) Magnitude response lines (dashed line – Simulated, solid line – Measured), (b) Phase response.

Figure 3.40 Photograph of the fabricated 0 dB coupled S-MS lines-based forward-wave directional coupler (FWDC).

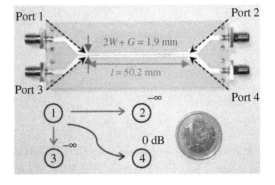

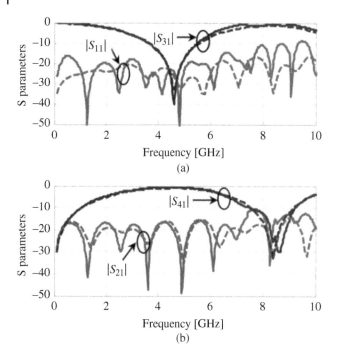

Figure 3.41 Performance of the fabricated 0-dB slow-wave FWDC, operating at 4.5 GHz (a) Return loss and isolation, (b) Coupling and insertion loss. (— measurements, ---simulations).

and ε^o are equal to 6 and 3.2, respectively. This leads to a coupling length l of 50.2 mm, in order to achieve a perfect cross-coupling that can be theoretically obtained when $S_{41} = 1$ and when the condition $l = \dfrac{C_0}{2f(\sqrt{\varepsilon^e} - \sqrt{\varepsilon^o})}$ is fulfilled (Bahl et al., 2007). It can be outlined that the length of a 0-dB matched FWDC designed on the same substrate without vias (i.e. using standard microstrip lines) would be $l = 320$ mm. It means that the proposed coupled S-MS lines-based FWDC enables an 84% length reduction as compared to a standard FWDC.

Let us remember that, ideally, a FWDC should present equal characteristic impedances $Z^e = Z^o = 50\ \Omega$ (Bahl et al., 2007). On the basis of Fig. 3.18(a), this condition could be reached for $G \simeq W \simeq 1$ mm. However, as shown in Fig. 3.18(b), a reduced difference in the square-root of the effective dielectric constants would be obtained. Hence, a compromise must be made between return loss and compactness.

The fabricated slow-wave FWDC was measured with a VNA Anritsu 37369A, the four ports of the coupler being connected to SMA connectors through four 50 Ω microstrip lines. After Short-Open-Load-Through (SOLT) calibration, a standard short/open de-embedding technique was applied at each port to remove the feeding lines influence. As shown in Fig. 3.41, measurement and simulation results are in good agreement.

At the operating frequency of 4.5 GHz, the measured insertion loss is 0.56 dB, the isolation being better than 18 dB at port 2 and 27 dB at port 3. The main limitation in terms of bandwidth comes from the isolation at port 3. By considering 10-dB isolation at port 3, the bandwidth extends from 3.6 to 5.6 GHz, corresponding to a fractional bandwidth (FBW) of 43%.

3.6.4 MEMS Phase Shifter With Liquid Crystal

The suspended S-MS lines presented in "Air-Filled Slow-Wave Microstrip Line AF-S-MS" were applied to design a slow-wave phase shifter using liquid crystal (LC) and MEMS technologies (Jost et al., 2018). It was developed using MnM technology. Nanowires grown in the porous alumina membrane under the suspended regions are mandatory to realize S-MS lines, but also in the context of liquid crystal phase shifters designed to concentrate the electric field in the LC-filled region, increasing the phase shift, and to be used as DC electrode for biasing the LC and actuating the MEMS bridge.

The use of MEMS with LC leads to increased phase shift and reduced pull-in voltage used to actuate the MEMS. Figure 3.42 illustrates two segments of the proposed phase shifter.

It is based on a S-MS line and suspended bridge-like segments filled with LC. The LC is used for an analog control of the phase shift, which is related to the alignment of the LC molecules with an applied DC electric field. Also, LC is used to reduce the MEMS actuation voltage, since the pull-in voltage is inversely proportional to the square-root of the relative dielectric constant (ε_r) of the medium, which is between 2.4 and 3.2 for the LC.

The phase shifter was designed based on 3D EM simulations. The nanowires were simulated as an equivalent anisotropic material with an equivalent conductivity of $\sigma = 1$ MS/m in the vertical direction and zero in the other directions. The nanoporous alumina substrate has a relative dielectric constant of 6.7 and loss tangent of 0.015 at 60 GHz and was simulated as a homogeneous material. The gap height and length L of each segment were

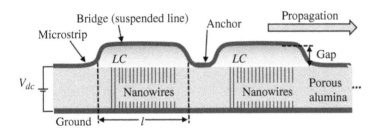

Figure 3.42 Longitudinal view of two segments of the phase shifter.

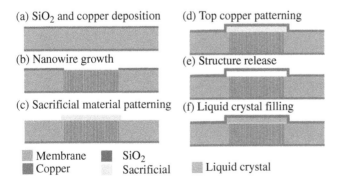

Figure 3.43 Fabrication process of the LC-filled S-MS MEMS phase shifter: (a) SiO$_2$ and copper deposition, (b) nanowire growth, (c) sacrificial material patterning, (d) top copper patterning, (e) structure release, and (f) liquid crystal filling.

determined in order to keep the pull-in voltage in air under 100 V (measurement equipment limit). A gap of 4 µm and a length of 500 µm were defined. The signal strip width was designed to keep the return loss better than 10 dB with the structure filled with LC. The steps of fabrication in a longitudinal view are presented in Fig. 3.43.

A 2350-µm-long phase shifter with four suspended segments separated by 50-µm-long anchoring regions was fabricated and measured. The LC GT3-23001 mixture from Merck KGaA was applied under the suspended regions with a micropipette. This LC has the following electric characteristics at 30 GHz: $2.47 \leq \varepsilon_r \leq 3.16$ and $15 \cdot 10^{-3} \geq \tan \delta \geq 3.3 \cdot 10^{-3}$.

In order to compare the different actuation principles, a phase shifter with a single suspended segment with $L = 500$ µm was also fabricated and tested under three different conditions: MEMS only (in air) switching at a biasing voltage of 90 V; LC only at a biasing voltage of 20 V; and a combination of LC and MEMS. In Fig. 3.45 it can be noted that the LC has the effect of lowering the pull-in voltage for the electrostatic switching, reducing the required 90 V for MEMS to 50 V for MEMS with LC. This combination also increases the total phase shift from 8° for MEMS or LC alone to 20° for MEMS with LC.

Figure 3.44 shows the fabrication results of the phase shifter comprising four suspended segments and the nanowire-filled regions.

The introduction of LC increases slightly the figure of merit (FoM), defined as the ratio between the maximum phase shift and the maximum insertion loss in dB, up to more than 40 GHz, when compared to MEMS, but considerably compared to LC alone, as also shown in Fig. 3.45.

Another inherent advantage of having the LC and MEMS combined is the capacity to implement fine, analog phase shift using the LC under low bias and coarse, digital phase shift at higher bias. Moreover, the individual bits of the digital phase shift can be implemented by changing the length L of each suspended section, thus changing their individual pull-in voltages. This effect can be seen in Fig. 3.46, where the phase shift of a four-segment-long phase shifter is plotted for different biasing. The analog tuning happens for bias voltages between 0 and 25 V; above this point, the individual segments start to switch, creating larger, discrete phase shifts.

Figure 3.47 shows the return and insertion loss for the complete phase shifter at zero biasing (black lines) and 50 V (gray lines). The insertion loss is better than 2 dB up to 40 GHz,

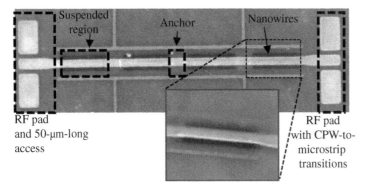

Figure 3.44 Fabricated phase shifter with four suspended segments showing in detail the suspended region and the nanowire-filled regions.

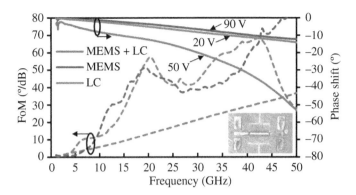

Figure 3.45 Influence in figure of merit (FoM) and phase shift of LC in S-MS MEMS phase shifter: one segment with $L = 500\,\mu m$ in air (MEMS) and with LC under the MEMS structure (MEMS+LC). In detail, 1-segment structure.

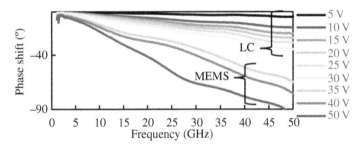

Figure 3.46 Analog and digital phase shifting of the S-MS phase shifter with MEMS and LC.

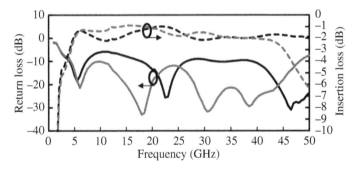

Figure 3.47 Insertion and return loss for a four-segment, each having $L = 500\,\mu m$ at zero biasing (black) and at 50 V (gray).

while the return loss is better than 10 dB. The presented results were de-embedded by software using a measured pad model.

Nanowire vias are used at the pads to realize a CPW-to-microstrip transition for probe measurement. A thin SiO_2 film creates a capacitive isolation between the CPW ground pads and the microstrip ground, thus decoupling the RF and DC grounds. For this reason, at frequencies below 5 GHz, there is no signal transmission. Above 45 GHz, when the

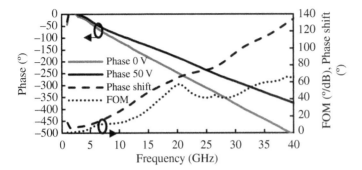

Figure 3.48 Phase at 0 and 50 V, total phase shift and FoM for the 4-segment S-MS MEMS phase shifter.

MEMS structure is actuated with 50 V, the insertion loss increases rapidly because of the low impedance (anchor region without nanowires) and high impedance (suspended region with nanowires) periodic structure, which results in a Bragg cut-off frequency. A bias-tee (SHF BT65R – HVC100) was used at the VNA ports (N5227B – Keysight) to apply DC to the S-MS line.

Figure 3.48 shows the phase, the FoM and the total phase shift of the 4-segment device. A total phase shift of 72° and 135° and a FoM of 42°/dB and 66°/dB were obtained at 24 and 40 GHz, respectively. The footprint of the device, without RF pads, is 0.13 mm² (2.35 mm × 55 µm).

3.7 CMOS Technology

In the same way as it was presented in the previous sections, the concept of S-MS line can be implemented in CMOS technology. The examples that are given in the following sections were carried out on the back-end-of-line (BEOL) of a 22-nm CMOS technology, but the approach is of course compatible with any type of BEOL, whether for CMOS or BiCMOS technologies.

In Section 3.7.1, the principle of an S-MS line in CMOS technology is presented and demonstrated experimentally, and the key design parameters are discussed. Next, the realization of an artificial transmission line made with L and C sections based on S-MS lines is described in Sections 3.7.2 and 3.7.3. Finally, Section 0 presents the realization of a branch-line coupler, made from S-MS lines, with the main objective of miniaturization, for an operating frequency of 60 GHz.

3.7.1 Slow-Wave Microstrip Lines (S-MS)

The design described in this section has been developed using the BEOL of the 22 nm FDSOI (Fully-Depleted Silicon-on-Insulator) CMOS process by Global Foundries. This technology provides 11 metal layers, as shown in Fig. 3.49.

For the signal strip, the topmost Alucap metal layer (LB in Fig. 3.49) was employed. The ground plane was realized by stacking two intermediate layers. Although preferable,

Figure 3.49 CMOS 22-nm Global Foundries back-end-of-line (BEOL).

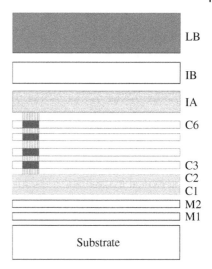

LB

IB

IA

C6

C3
C2
C1
M2
M1

Substrate

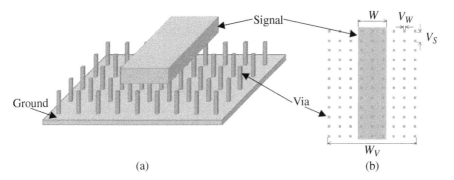

Signal

Ground

Via

(a)

W V_W

V_S

W_V

(b)

Figure 3.50 Monolithically integrated S-MS line based on bed of nails: (a) 3D view, (b) top view.

the choice of lower metal layers, i.e. M1 and M2, was not possible due to stringent density design rules posed by the selected technology. Hence, the two metals above M1 and M2 were chosen and stacked together, i.e. C1 and C2. To achieve the slow-wave effect, an array of vias was used by stacking intermediate metallic layers connected to the ground plane. The distance between the signal strip and the ground plane is 4.3 µm, and the height of the vias is 2.185 µm.

As shown in Fig. 3.50, a signal strip of width W is made on a lattice of square vias, with side V_W and periodicity V_S. Indeed, as explained in Section 3.2, it has to be ensured that the electrical field does not reach the ground; electric field lines must be captured by the vias. At the same time, a minimum of vias must be used, not to increase the losses due to the eddy currents, in particular when dealing with mm-waves. In order to extend this behavior to the fringing fields, the width of the array of vias, W_V, should be wider than the signal strip.

The performance of the S-MS lines with characteristic impedance Z_0 equal to 35 and 50 Ω was evaluated through EM simulations and compared with the ones of two conventional microstrip lines. In the 35-Ω case, the signal strip width, W, is equal to 11 µm for the microstrip line and 9 μm for the S-MS line, respectively. In the 50-Ω case, these strips' widths equal 7.25 and 4.5 μm, respectively. The narrower strips obtained for S-MS lines are interesting, because they allow, in the case of meandered lines, to significantly reduce the occupied surface. This is shown in Section 0, where a branch-line coupler is designed with an improved overall compactness. In both slow-wave configurations (50 and 35 Ω), the array of vias width, W_V, is the same, for simplicity; it is equal to 18 μm, which is enough to capture the fringing field. The lattice parameters are $V_W = 1.2$ μm and $V_S = 3$ μm.

The comparison between the 50-Ω microstrip lines and S-MS lines is given in Figs. 3.51 and 3.52, respectively.

At 60 GHz, the effective dielectric constant $\varepsilon_{r_{eff}}$ of the 50-Ω S-MS line is equal to 4.65, whereas for the conventional microstrip line $\varepsilon_{r_{eff}}$ is equal to 3.88, thus resulting in a SWF of 1.1. The miniaturization is obtained at the expense of an attenuation constant α that is higher for the S-MS line. The increased losses are mainly due to the currents inside the vias. Even so, the Q-factors of the two transmission lines' structures are similar, as the increase in the phase constant β compensates the increase in the attenuation constant α. A similar behavior occurs for the 35-Ω characteristic impedance microstrip and S-MS lines.

Thus, the S-MS line is interesting for making passive circuits designed from transmission lines of fixed electrical length, often quarter-wave transmission lines. The losses are the

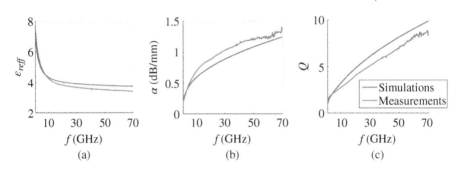

Figure 3.51 Microstrip line extracted (a) effective relative permittivity $\varepsilon_{r_{eff}}$, (b) attenuation constant α, and (c) quality factor Q.

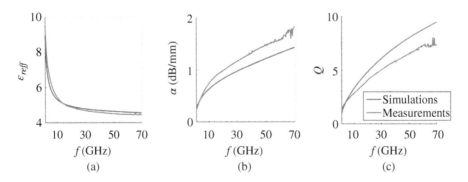

Figure 3.52 S-MS line extracted $\varepsilon_{r_{eff}}$, α and Q.

same for a given electrical length, but with a lower physical length, and, as underlined above, the width is lower, giving more latitude to make meandered lines.

3.7.2 Principle of an Artificial Transmission Line Based on Meandered S-MS Lines

The form factor of the transmission lines can be an issue, with widths of a few μm and lengths of up to a mm. Also, it is often interesting to meander the strips in order to approach a generally square shape. We show in this subsection that mixing meandered-microstrip lines (m-MS) and S-MS lines used as open stubs can be a very good option that leads to efficient transmission lines, both in terms of footprint and electrical performance. The proposed concept, called meandered-MicroStrip line with Slow-wave open Stub (m-MS-SoS) (Acri et al., 2019), is presented in Fig. 3.53. The m-MS line (Fig. 3.53[a]) acts as an inductance L. It is loaded by a slow-wave open stub (SoS) acting as a capacitive load C (Fig. 3.53[b]). This leads to an artificial transmission line configuration made of L and C sections.

The microstrip line is meandering to achieve a more acceptable form factor. In the next sections, the proposed artificial transmission line is compared to a simple m-MS line and design insights and simulated results are given.

3.7.3 Artificial S-MS Line and Meandered-Microstrip Line

3.7.3.1 Design
Each $L - C$ cell of the artificial transmission line is now utilized as a quarter-wavelength 50-Ω branch, that could be employed for several purposes, in particular for mm-wave branch-line couplers, WPDs/combiners, filters, etc. In order to achieve a quarter-wavelength, four cells were employed, each cell providing a $\lambda_g/16$ length, where λ_g is the guided wavelength, at a working frequency of 60 GHz. One of the most important

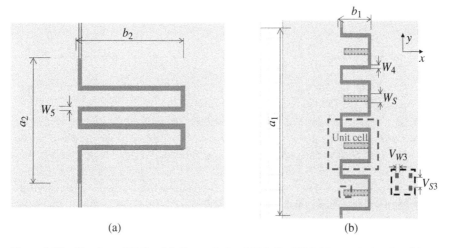

(a)   (b)

Figure 3.53 Simple m-MS line (a); Concept of m-MS-SoS artificial transmission line (b).

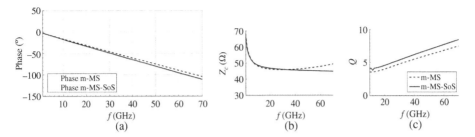

Figure 3.54 Electrical length (a), characteristic impedance (b), and Q-factor (c) comparison between S-MS artificial transmission line (m-MS-SoS) and meandered MS (m-MS).

features of this transmission line is its capability to keep constant Z_c, while increasing β. This is possible since L and C are synthesized independently; hence, they can be increased at the same time, thus lowering the phase velocity given by $v_\varphi = \frac{1}{\sqrt{LC}}$ and, consequently, reducing the physical length l for a given electrical length θ, since $\theta = \frac{\omega}{v_\varphi} l$. To get $Z_c = 50\ \Omega$ and $\theta = 90°$, the m-MS-SoS and m-MS lines were sized as follows: W_4, W_S, a_1, b_1, V_{W3} and V_{S3} (see Fig. 3.53[a]) were set equal to 4, 8, 248, 54.5, 1.2, and 4.6 μm, respectively, and W_5, a_2 and b_2 (see Fig. 3.53[b]) were set to 4, 126, and 144 μm, respectively.

3.7.3.2 Results and Comparison

The m-MS-SoS line and the m-MS line are compared in terms of Z_c and Q in Fig. 3.54. For the same electrical length (Fig. 3.54[a]), and for almost the same value of Z_c (Fig. 3.54[b]), the Q-factor of the m-MS-SoS line (Fig. 3.54[c]) is higher than its m-MS line counterpart, i.e. 7.8 as compared to 6.8.

In order to also compare the area efficiency of the artificial transmission line, an area factor (AF) was defined as the ratio between the electrical length and the surface area:

$$AF = \left(\frac{Electrical\ length(\theta)}{Surface\ area} \right) [rad/(mm^2)] \tag{3.37}$$

The AF is equal to $72.7 \frac{rad}{mm^2}$ for the m-MS line and $82.5 \frac{rad}{mm^2}$ for the m-MS-SoS line, respectively. Thus, the m-MS-SoS line is slightly more compact.

In order to highlight some trade-offs in designing transmission lines, the performance of the transmission lines presented above is summarized in Table 3.4. The m-MS-SoS line exhibits the highest SWF, with a Q-factor that is comparable to the classical microstrip line. It can be noticed that the AF of the proposed m-MS-SoS line is lower than the one of the S-MS line. However, if the example of a branch-line coupler or power divider is considered, it is mandatory to meander the S-MS lines, leading to an increase in the surface area. Hence, in practical applications, the proposed m-MS-SoS line offers a nice trade-off between occupied area and electrical performance, as illustrated in the next section dedicated to the design of a branch-line coupler using m-MS-SoS lines.

Table 3.4 Transmission lines overall performance.

		Extracted performance for $Z_c = 50\ \Omega$			
	$\varepsilon_{r_{eff}}$	α (dB/mm)	Q	AF (rad/mm²)	SWF
MS	3.45	1.2	8.3	68.1	1
S-MS	4.47	1.6	7.3	84.3	1.14
m-MS	3.28	1.44	6.8	72.7	~1
m-MS-SoS	**28.5**	**3.75**	**7.8**	**82.5**	**2.9**

3.7.4 Branch-Line Coupler

3.7.4.1 Design

A branch-line coupler was designed at 60 GHz to prove the miniaturization performance achievable employing the technique described in Section 3.7.3. As it is well known, a conventional branch-line coupler is made up of four quarter-wavelength transmission lines, as shown in Fig. 3.55, with characteristic impedances equal to $Z_c = 50\ \Omega$ and $Z_c/\sqrt{2} = 35\ \Omega$ when 50-Ω ports impedance is considered.

To begin, the two transmission lines must be synthesized, with characteristic impedance Z equal to 50 and 35 Ω, respectively. The results in Table 3.4 show that transmission lines of the S-MS and m-MS-SoS type are good candidates for the synthesis of these two transmission lines (Fig. 3.56), hence meandered S-MS and m-MS-SoS were considered to realize the branch-line coupler.

Four cells were employed to build a quarter-wavelength transmission line with the m-MS-SoS line type. More cells could be employed, but cells of length equal to $\frac{\lambda_g}{16}$ are short enough to consider a distributed transmission line type. The geometrical parameters of the two transmission lines are reported in Table 3.5.

Next, different form factors can be achieved for each branch of the coupler. This degree of freedom can be exploited to optimize the miniaturization of the branch-line coupler. The four most interesting combinations were simulated and compared in terms of area and insertion loss (Fig. 3.57).

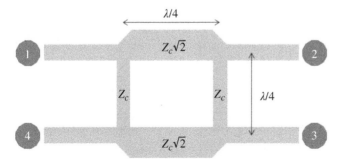

Figure 3.55 Branch-line coupler and ports definition.

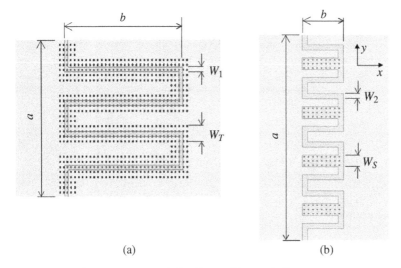

(a)   (b)

Figure 3.56 Slow-wave transmission lines utilized to build the branch-line coupler. m-S-MS line (a), and m-MS-SoS line (b).

Table 3.5 Geometrical parameters of the two quarter-wavelength transmission lines synthesized with S-MS and m-MS-SoS lines type (all dimensions in µm).

Z Name	35 Ω		50 Ω	
	m-S-MS	m-MS-SoS	m-S-MS	m-MS-SoS
W_1	8	—	4	—
W_T	18	—	8	—
W_2	—	7	—	3.5
W_S	—	16.3	—	8

The four configurations were obtained by combining the two types of quarter-wavelength transmission line sections, as reported in Fig. 3.57. Although the performance of the four configurations in Fig. 3.57 is very similar, with insertion loss between 2.2 and 2.6 dB, the resulting surface areas vary significantly. The smallest size is achieved for "Configuration 2" when using the m-MS-SoS and the m-S-MS for the 35 and 50-Ω branches, respectively.

3.7.4.2 Results

The branch-line coupler in "Configuration 2" was measured using a 145-GHz 4 ports VNA (ANRITSU ME7838D-4), along with on-wafer probing (GGB Picoprobe 145A-GSGSG-50-P-N, dual probes, 50-µm pitch, nickel-alloy tips). The layout, the measured S-parameters, along with those generated by 3D EM simulations, are given in Fig. 3.58.

The measured insertion loss between the input and the direct and coupled ports is 1.55 and 2.35 dB, respectively, at 60 GHz. The isolation is higher than 20 dB, and the return loss

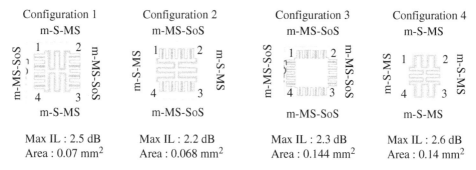

Max IL : 2.5 dB
Area : 0.07 mm²

Max IL : 2.2 dB
Area : 0.068 mm²

Max IL : 2.3 dB
Area : 0.144 mm²

Max IL : 2.6 dB
Area : 0.14 mm²

Figure 3.57 Comparison of four different configurations of branch-line coupler realized using m-S-MS and m-MS-SoS quarter-wave line sections. Insertion loss (IL) and area are given below each configuration.

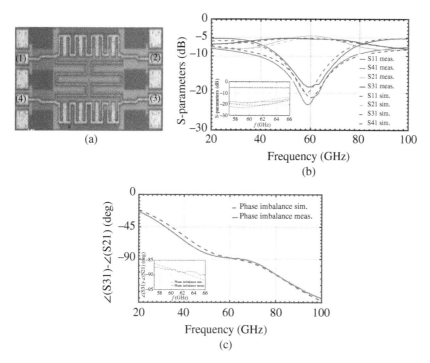

Figure 3.58 Branch-line coupler layout (a), S-parameters (b) and Phase imbalance (c).

equals 18 dB. Measurements showed an amplitude imbalance of 0.8 dB, slightly higher than the 0.3 dB provided by simulation. Nevertheless, measurement and simulation results are in good agreement all over the bandwidth. The measured phase difference between the direct and coupled ports is 88.2°, which is only 1.8° off from the ideal case (Fig. 3.58[b]). All ports report a return loss higher than 15 dB over the unlicensed band 54–66 GHz. The configuration presented in this section exhibits an interesting trade-off between miniaturization concerns and electrical performance.

Table 3.6 Comparison of expected insertion loss and *Q*-factors at 60 GHz with different BEOL thicknesses for a microstrip line.

Substrate thickness (μm)	Insertion loss (dB/mm)	Q-factor
10	0.34	28
7	0.46	20.4
4.5	0.7	13.6

3.7.4.3 Influence of the Back-End-Of-Line

Performance is not only influenced by the miniaturization approach but also by the BEOL of the considered technology. In this section, it will be demonstrated that the technology rather than the miniaturization methodology has a great influence on the performance of the proposed S-MS meandered lines and thus on the performance of the coupler designed in the previous section.

Indeed, the thinner the BEOL (substrate), the narrower the signal strip of the microstrip line for a given characteristic impedance, and hence the higher its series resistance and its attenuation constant. To have a quantitative assessment of this effect, a conventional 50-Ω microstrip line was designed and simulated for several BEOL thicknesses. For comparison purposes, the substrate dielectric constant was fixed at 4, which is a realistic value for CMOS technologies. Moreover, metal layers were designed in copper with signal strips and ground thicknesses of 2 and 0.25 μm, respectively. Results are summarized in Table 3.6, for a working frequency equal to 60 GHz.

The higher the substrate thickness, the higher the *Q*-factor. These results mean that the technique proposed in this book would give better results, in terms of insertion loss, if applied to a thicker BEOL like in (Ding et al., 2007) (more than 10 μm). To emphasize these simple results, the coupler designed in Section 3.7.4.1 was simulated considering a thick substrate of about 10 μm (Zimmer et al., 2021). The simulated results show that a branch-line coupler designed employing the proposed miniaturization technique would exhibit an insertion loss of 0.6 dB, thus showing the importance of the BEOL when dealing with mm-wave passive circuits performance.

References

Abouchahine, S., Alhalabi, H., Baheti, A., Ferrari, P., Issa, H., Kaddour, D., Pistono, E., & Podevin, F. (2018). Miniaturized branch-line coupler based on slow-wave microstrip lines. *International Journal of Microwave and Wireless Technologies*, 10(10), 1103–1106. https://doi.org/10.1017/S1759078718001204

Acri, G., Podevin, F., Pistono, E., Boccia, L., Corrao, N., Lim, T., Isa, E. N., & Ferrari, P. (2019). A millimeter-wave miniature branch-line coupler in 22-nm CMOS technology. *IEEE Solid-State Circuits Letters*, 2(6), 45–48. https://doi.org/10.1109/LSSC.2019.2930197

Bahl I. J., Bhartia P., Hong J. R., & Mongia R.K.2007). *RF and Microwave Coupled line Circuits*. Artech House, 123–161.

Blair, D. G., Woode, R. A., Tobar, M. E., & Ivanov, E. N. (1994). Measurement of dielectric loss tangent of alumina at microwave frequencies and room temperature. *Electronics Letters*, 30(25), 2120–2122. https://doi.org/10.1049/el:19941470

Burdin, F., Podevin, F., & Ferrari, P. (2016). Flexible and miniaturized power divider. *International Journal of Microwave and Wireless Technologies*, 8(3), 547–557. https://doi.org/10.1017/S1759078715000252

Cathelin, A., Martineau, B., Seller, N., Douyere, S., Gorisse, J., Pruvost, S., Raynaud, Ch., Gianesello, F., Montusclat, S., Voinigescu, S. P., Niknejad, A. M., Belot, D., & Schoellkopf, J. P. (2007). Design for millimeter-wave applications in silicon technologies. *ESSCIRC 2007 – 33rd European Solid-State Circuits Conference*, 464–471. https://doi.org/10.1109/ESSCIRC.2007.4430343

Coulombe, M., Nguyen, H. V., & Caloz, C. (2007). Substrate integrated artificial dielectric (SIAD) structure for miniaturized microstrip circuits. *IEEE Antennas and Wireless Propagation Letters*, 6, 575–579. https://doi.org/10.1109/LAWP.2007.910959

Ding, H., Wang, G., Lam, K., Jordan, D., Zeeb, A., Bavisi, A., Mina, E., & Gaucher, B. (2007). Modeling and implementation of on-chip millimeter-wave compact branch line couplers in a BiCMOS technology. *European Microwave Conference*, 2007, 458–461. https://doi.org/10.1109/EUMC.2007.4405226

Dubuc, D., Grenier, K., Fujita, H., & Toshiyoshi, H. (2009a). Plastic-based microfabrication of artificial dielectric for miniaturized microwave integrated circuits. *Metamaterials*, 3(3–4), 165–173. https://doi.org/10.1016/j.metmat.2009.09.002

Dubuc, D., Grenier, K., Fujita, H., & Toshiyoshi, H. (2009b). Micro-fabricated tunable artificial dielectric for reconfigurable microwave circuits. *European Microwave Week 2009, EuMW 2009: Science, Progress and Quality at Radiofrequencies, Conference Proceedings – 39th European Microwave Conference, EuMC 2009*. https://doi.org/10.1109/EUMC.2009.5296366

Evans, R. J., Skafidas, E., & Yang, B. (2012). Slow-wave slot microstrip transmission line and bandpass filter for compact millimetre-wave integrated circuits on bulk complementary metal oxide semiconductor. *IET Microwaves, Antennas & Propagation*, 6(14), 1548–1555. https://doi.org/10.1049/iet-map.2012.0336

Gammand, P., & Bajon, D. (1990). Slow wave propagation using interconnections for IC technologies. *Electronics Letters*. https://doi.org/10.1049/el:19900926

Hammerstad, E., & Jensen, O. (1980). Accurate models for microstrip computer-aided design. *MTT-S International Microwave Symposium Digest*, 80(1), 407–409. https://doi.org/10.1109/MWSYM.1980.1124303

Jost, M., Gautam, J. S. K., Gomes, L. G., Reese, R., Polat, E., Nickel, M., Pinheiro, J. M., Serrano, A. L. C., Maune, H., Rehder, G. P., Ferrari, P., & Jakoby, R. (2018). Miniaturized liquid crystal slow wave phase shifter based on nanowire filled membranes. *IEEE Microwave and Wireless Components Letters*, 28(8). https://doi.org/10.1109/LMWC.2018.2845938

Lee, J. J., & Park, C. S. (2010). A slow-wave microstrip line with a high-Q and a high dielectric constant for millimeter-wave CMOS application. *IEEE Microwave and Wireless Components Letters*, 20, 381–383. https://doi.org/10.1109/LMWC.2010.2049430

Li, D., Gao, J., Austermann, J. E., Beall, J. A., Becker, D., Cho, H.-M., Fox, A. E., Halverson, N., Henning, J., Hilton, G. C., Hubmayr, J., Irwin, K. D., Van Lanen, J., Nibarger, J., & Niemack, M. (2013). Improvements in silicon oxide dielectric loss for superconducting microwave detector circuits. *IEEE Transactions on Applied Superconductivity*, 23(3), 1501204. https://doi.org/10.1109/TASC.2013.2242951

Li, L., Xu, F., Wu, K., Delprat, S., & Chakef, M. (2005). Slow-wave line filter design using full-wave circuit model of interdigital capacitor. *2005 European Microwave Conference, 2005, Paris, France*, 4. https://doi.org/10.1109/EUMC.2005.1608880

Luong, D., Acri, G., Podevin, F., Vincent, D., Pistono, E., Serrano, A., & Ferrari, P. (2019). Forward-wave directional coupler based on slow-wave coupled microstrip lines. *IET Microwaves, Antennas & Propagation*, 13(14), 2486–2489. https://doi.org/10.1049/iet-map .2019.0296

Machac, J. (2006). Microstrip line on an artificial dielectric substrate. *IEEE Microwave and Wireless Components Letters*, 16(7), 416–418. https://doi.org/10.1109/LMWC.2006.877120

Mangan, A. M., Voinigescu, S. P., Yang, M.-T., & Tazlauanu, M. (2006). De-embedding transmission line measurements for accurate modeling of IC designs. *IEEE Transactions on Electron Devices*, 53(2), 235–241. https://doi.org/10.1109/TED.2005.861726

Marks, R. B., & Williams, D. F. (1991). Characteristic impedance determination using propagation constant measurement. *IEEE Microwave and Guided Wave Letters*, 1(6), 141–143. https://doi.org/10.1109/75.91092

Masuda, H., & Fukuda, K. (1995). Ordered metal nanohole arrays made by a two-step replication of honeycomb structures of anodic alumina. *Science*, 268(5216), 1466–1468. https://doi.org/10.1126/science.268.5216.1466

Paul, C. R. (2010). *Inductance: Loop and Partial*. John Wiley & Sons.

Rautio, J. C. (2000). An investigation of microstrip conductor loss. *IEEE Microwave Magazine*, 1(4), 60–67. https://doi.org/10.1109/6668.893247

Serrano, A. L. C., Franc, A. L., Assis, D. P., Podevin, F., Rehder, G. P., Corrao, N., & Ferrari, P. (2014a). Slow-wave microstrip line on nanowire-based alumina membrane. *IEEE MTT-S International Microwave Symposium Digest* (IMS2014)*, 2014, Tampa, FL, USA*, 1–4. https:// doi.org/10.1109/MWSYM.2014.6848552

Serrano, A. L. C., Franc, A.-L., Assis, D. P., Podevin, F., Rehder, G. P., Corrao, N., & Ferrari, P. (2014b). Modeling and characterization of slow-wave microstrip lines on metallic-nanowire-filled-membrane substrate. *IEEE Transactions on Microwave Theory and Techniques*, 62(12), 3249–3254. https://doi.org/10.1109/TMTT.2014.2366108

Wang, G., Woods, W., Ding, H., & Mina, E. (2009). Novel on-chip high performance slow wave structure using discontinuous microstrip lines and multi-layer ground for compact millimeter wave applications. *2009 59th Electronic Components and Technology Conference*, 2009, San Diego, CA, USA, 1606–1611. https://doi.org/10.1109/ECTC.2009.5074229

Wu, C.-K., Wu, H.-S., & Tzuang, C.-K. C. (2002). Electric-magnetic-electric slow-wave microstrip line and bandpass filter of compressed size. *IEEE Transactions on Microwave Theory and Techniques*, 50(8), 1996–2004. https://doi.org/10.1109/TMTT.2002.801355

Yang, F. R., Qian, Y., Coccioli, R., & Itoh, T. (1998). A novel low-loss slow-wave microstrip structure. *IEEE Microwave and Guided Wave Letters*, 8(11), 372–374. https://doi.org/10.1109/ 75.736247

Zhao, W., Li, X., Gu, S., Kang, S. H., Nowak, M. M., & Cao, Y. (2009). Field-based capacitance modeling for sub-65-nm on-chip interconnect. *IEEE Transactions on Electron Devices*, 56(9), 1862–1872. https://doi.org/10.1109/TED.2009.2026162

Zimmer, T., et al. (2021). SiGe HBTs and BiCMOS technology for present and future millimeter-wave systems. *IEEE Journal of Microwaves*, 1(1), 288–298. https://doi.org/10 .1109/JMW.2020.3031831

4

Slow-Wave SIW

Matthieu Bertrand[1], Jordan Corsi[2], Emmanuel Pistono[2], and Gustavo P. Rehder[3]

[1] *IMEP-LAHC, CNRS, Grenoble INP, Université Grenoble Alpes, Grenoble, France*
[2] *TIMA, Université Grenoble Alpes, CNRS, Grenoble INP, Grenoble, France*
[3] *Polytechnic School, University of São Paulo, São Paulo, Brazil*

This chapter focuses on slow-wave Substrate Integrated Waveguides (SW-SIW), which were demonstrated in 2014 (Niembro-Martin et al., 2014). The development of Substrate Integrated Waveguides (SIWs) themselves is relatively recent, with publications dating back to late 1990s and early 2000s, as for example in (Uchimura et al., 1998) and (Deslandes & Wu, 2003). To assess the specific advantages of SIWs for RF circuit implementation, it is important to compare them to traditional technologies: conventional metallic waveguides and planar transmission lines, particularly microstrip lines. Compared to conventional metallic waveguides, SIWs are significantly less expensive due to the use of planar technologies compatible with conventional Printed Circuit Board (PCB) processes. The interest of SIWs with respect to microstrip lines lies in three characteristics, (i) the electrical performance, i.e. insertion loss and unloaded quality factor, offering the possibility of realizing resonant circuits (filters in particular) more efficiently than their counterparts made in conventional low-profile technologies (quality factor comprised between 150 and 1000 depending on the substrate loss tangent); (ii) the electromagnetic immunity against external parasitic radiation due to their closed waveguide structure; and (iii) high power capability as compared to microstrip lines. However, like conventional waveguides, their main drawback is their lateral dimension, which is of the order of half a wavelength for the first propagation mode. Therefore, developing methods to reduce the dimensions of SIWs is of great interest.

The use of the slow-wave effect, in a similar manner as used in Chapter 3 for microstrip lines, is one of the current solutions to achieve a reduction in the dimensions of SIWs. The application of this concept to SIWs is the subject of this chapter.

The chapter is organized as follows. In the first section, the principle of SIW and slow-wave Substrate Integrated Wave Guide (SW-SIW) is presented. Then, the modeling of the slow-wave SIW is presented for a lossless parallel plate waveguide. Afterwards, the dielectric and metallic losses are considered. The next two sections are devoted to the presentation of SW-SIWs in two very different technologies, namely the classic PCB technology and the MnM (for Metallic nanowire Membrane) technology based on a

Slow-Wave Microwave and mm-Wave Passive Circuits, First Edition. Edited by Philippe Ferrari, Anne-Laure Franc, Marc Margalef-Rovira, Gustavo P. Rehder, and Ariana Lacorte Caniato Serrano.
© 2025 John Wiley & Sons Ltd. Published 2025 by John Wiley & Sons Ltd.

nanoporous alumina substrate. Partially air-filled slow-wave SIWs (PAF-SW-SIW) are also featured in each technology.

4.1 Substrate Integrated Waveguides

Conventional Substrate Integrated Waveguides (SIW) are the transposition of bulky metallic waveguides in a low-profile form such as in PCB technology. Starting with a substrate, the top and bottom walls are formed by metal layers on either side of the substrate, the lateral walls being imitated by two rows of closely-spaced metallized vias. A 3D view of a SIW is given in Fig. 4.1. It is made of a dielectric substrate of height h, the waveguide being delimited by two rows of vias separated by a width W. The diameter of the vias is called d, and they are spaced by a distance S. ε and μ are the permittivity and permeability of the substrate, respectively.

SIWs are similar in construction as conventional rectangular waveguides apart from the lateral walls that are made of vias. In both cases, the propagation of electromagnetic waves, occurs only in non-TEM modes. For rectangular waveguide, only transverse electric (TE_{mn}) or transverse magnetic (TM_{mn}) propagation modes are possible, where m and n are the indices of the considered propagation mode. For SIWs, TM propagation modes cannot exist due to the lateral walls made of vias, prohibiting any longitudinal currents (Feng & Wu, 2005). In both metallic waveguides and SIWs, each of these propagation modes has a cut-off frequency, below which propagation is not possible. It is common to use rectangular waveguides in a frequency band where there is only one mode of propagation, namely in the frequency band of the first mode TE_{10}. The frequency band where there is only one mode of propagation is called the single-mode band. Generally, it is between f_{c10} and f_{c20} in the case of the TE_{10} mode, that is, between f_{c10} and $2f_{c10}$. In reality, the operating zone is often reduced to be between $1.2\,f_{c10}$ and $1.8\,f_{c10}$ to avoid frequency zones with high dispersion where the attenuation constant may be higher than its value in the middle of the band around $1.5\,f_{c10}$.

For rectangular waveguides, the cutoff frequency f_c of TE_{10} mode is expressed by equation (4.1) and it is dependent on the transverse dimensions of the guide, i.e. its width W and height h, as well as the permittivity ε and permeability μ of the dielectric material inside the rectangular guide (Pozar, 2011). The phase constant β_{TE} of the TE modes is described by the equation (4.2), where k represents the free-space wavenumber (equation (4.3)) for a given angular frequency ω and propagation velocity c in the substrate medium. Equation (4.4) represents the cut-off wavenumber k_c. The parameter β_{TE} is a real number when k^2 is greater than k_c^2, that is, for a frequency higher than the cutoff frequency. Otherwise, the phase constant is purely imaginary, which reflects the evanescent nature of the wave in the waveguide at these frequencies (Pozar, 2011). ε_r and μ_r correspond to the relative dielectric constant and permeability of the substrate, respectively.

Figure 4.1 SIW geometry (3–dimensional view).

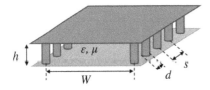

$$f_{c_{TE_{mn}}} = \frac{1}{2\pi\sqrt{\varepsilon\mu}}\sqrt{\left(\frac{m\pi}{W}\right)^2 + \left(\frac{n\pi}{h}\right)^2} \rightarrow f_c = \frac{c_0}{2W\sqrt{\varepsilon_r}} \tag{4.1}$$

$$\beta_{TE_{mn}} = \sqrt{k^2 - k_c^2} \tag{4.2}$$

$$k = \omega\sqrt{\varepsilon\mu} = \frac{\omega}{c} \tag{4.3}$$

$$k_c = \sqrt{\left(\frac{m\pi}{W}\right)^2 + \left(\frac{n\pi}{h}\right)^2} \tag{4.4}$$

$$c = \frac{1}{\sqrt{\varepsilon\mu}} = \frac{c_0}{\sqrt{\varepsilon_r\mu_r}} \tag{4.5}$$

In SIWs, the lateral rows of metallized vias behave like solid walls as long as certain design rules are respected to reduce power loss due to radiation between these vias (Feng & Wu, 2005). These rules were expressed in (Deslandes, 2005) and are recalled below, where λ_g, d, and s correspond to the guided wavelength, the diameter of the vias, and their center-to-center spacing, respectively (Fig. 4.1). Thus, according to expression (4.6), the higher the operating frequency desired, the smaller the spacing between the vias must be to avoid leakage of the electromagnetic wave through the side walls. Moreover, equation (4.7) indicates that the periodicity of the vias must be greater than the diameter of the vias for obvious geometrical reasons and must be less than twice this diameter to minimize radiation losses (Bozzi et al., 2007).

$$0.05\lambda_g < s < \frac{\lambda_g}{4} \tag{4.6}$$

$$d < s < 2d \tag{4.7}$$

The cutoff frequency of the TE_{10} mode in SIW is defined by (4.8) that is similar to (4.1), with W replaced by the effective width of the waveguide W_{eff}. In practice, the thickness h is much smaller than 2 times the width W, and the cut-off frequency f_c is independent of the substrate thickness. Equation (4.10) also expresses the phase constant β as function of W_{eff}.

$$f_{c\_SIW} = \frac{c_0}{2W_{eff}\sqrt{\varepsilon_r}} \tag{4.8}$$

$$\beta_{TE_{10}} = \sqrt{\left(\frac{\omega \cdot \sqrt{\varepsilon_r}}{c_0}\right)^2 - \left(\frac{\pi}{W_{eff}}\right)^2} \tag{4.9}$$

The W_{eff} parameter is used to account for the small penetration of the wave between the vias. In the literature, there are different expressions to determine W_{eff}. Equation (4.10) from (Cassivi et al., 2002) is sufficiently accurate if design rules (4.6) and (4.7) are followed, i.e. for values of d and s much smaller than λ_g. However, equation (4.11) from (Feng & Wu, 2005) can be used in the case of large vias or non-negligible d/W ratios. Finally, a third expression (4.12) from (Salehi & Mehrshahi, 2011), developed using an analytical method, allows for the calculation of W_{eff} over a wider range of d and s parameters than expressions (4.10) and (4.11).

$$W_{eff} = W - \frac{d^2}{0.95s} \tag{4.10}$$

$$W_{eff} = W - 1.08\frac{d^2}{s} + 0.1\frac{d^2}{W} \tag{4.11}$$

$$W_{eff} = \frac{W}{\sqrt{1 + \left(\frac{2W-d}{s}\right)\left(\frac{d}{W-d}\right)^2 - \frac{4W}{5s^4}\left(\frac{d^2}{W-d}\right)^3}} \tag{4.12}$$

4.2 Basic Concept of the Slow-Wave SIW

The slow-wave effect can be created by the spatial separation of the electric and magnetic energies, as shown in Chapter 1. Different topologies to create slow-wave SIW have been proposed. Two topologies are based on the idea of loading the top cover of the SIW either with polyline (Jin et al., 2016) or lumped SMD – *Surface Mounted Device* (Jin et al., 2017) inductors to create an additional inductive effect. They are both illustrated in Fig. 4.2.

For a given cut-off frequency and electrical length, the inductive loading brings a reduction between 35 and 40% in both lateral and longitudinal directions. The reduction in lateral and longitudinal directions can also be independently adjusted by changing the inductance values.

In a SIW, another possibility to achieve slow-wave propagation consists in confining the electric field to a region of the SIW, ideally without perturbing the magnetic field. This can be done using metallic posts inside the SIW, connecting only one metallic layer, as shown in Fig. 4.3. This approach was first presented by (Niembro-Martin et al., 2014) and is the focus of this chapter.

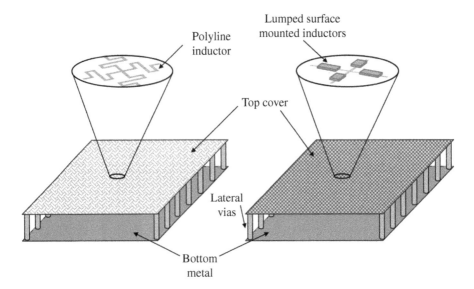

Figure 4.2 Slow-wave SIW based on inductive loading of the top cove. Left: polyline Source: Adapted from Jin et al., (2016). Right: Lumped surface mounted inductors. Source: Adapted from Jin et al., (2017).

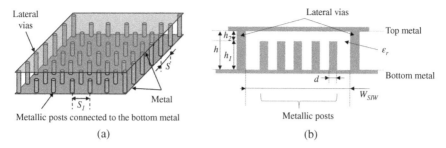

Figure 4.3 Schematic view of the SW-SIW. (a) 3–dimensional view, (b) transversal cross-section of the SW-SIW (example with 5 rows of metallic posts).

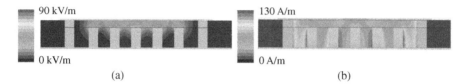

Figure 4.4 Cross-section view of the SW-SIW. (a) Electric field magnitude, (b) Magnetic field magnitude. ANSYS HFSS simulation. Source: Niembro-Martin et al., (2014)/with permission of IEEE.

Figure 4.3 shows a SIW of height h, with width W_{SIW} defined by two rows of periodic vias (lateral vias) connecting the top and bottom metallic layers on either side. Inside the SIW, a dense distribution of metallic posts of height h_1 and diameter d are used to capture the electric field and to confine it to the dielectric region of height h_2 of the SIW i.e. without metallic post.

The spatial separation between electric and magnetic fields can be illustrated with a numerical simulation of the SW-SIW in a commercial electromagnetic tool as shown in Fig. 4.4. Figure 4.4(a) shows that the electric field is mainly concentrated between the top of the metallic posts and the top metal layer, whereas the magnetic field magnitude (Fig. 4.4(b)) follows the typical SIW distribution, with the field flowing around the metallic posts. This behavior is a typical phenomenon for slow-wave transmission lines, which require a separation of electric and magnetic fields (Bertrand et al., 2020).

4.3 Modeling of Slow-Wave SIW

Parallel Plate Waveguides (PPW) can be used to model the slow-wave effect, effective dielectric constant and cutoff frequency of the SW-SIWs. First, a lossless case is derived for a Slow-wave PPW (SW-PPW). Then, the effect of the dielectric loss and the conductivity of the metallic posts is considered. In both cases (with or without losses), we derive the model for SW-SIW from the analysis of the SW-PPW.

4.3.1 Lossless SW-PPW to Lossless SW-SIW

Here we consider an ideal scenario to relate velocity reduction and fields spatial separation and derive the slow-wave factor for lossless PPW, as proposed in (Bertrand et al., 2020).

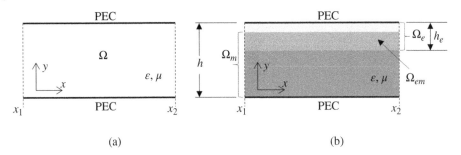

Figure 4.5 Transverse section of both (a) reference and (b) slow-wave parallel plate waveguide (PPWs) for derivation of the SWF. A Perfect Electric Conductor (PEC) is used in both cases for the parallel plates.

First of all, the reference waveguide (see Fig. 4.5(a)) is a lossless PPW filled with a medium of dielectric constant ε and magnetic permeability μ. It is assumed that its height is lower than half the free-space wavelength in the considered medium, so that only the fundamental transverse electromagnetic mode can propagate between the two perfectly conducting planes.

Based on these assumptions, it follows that the power flux is distributed homogeneously in the transverse section. The electric and magnetic field vectors for the TEM propagation mode in this structure are denoted $\boldsymbol{E}_{ref}$ and $\boldsymbol{H}_{ref}$. Considering propagation along the z axis, it follows that $\boldsymbol{E}_{ref}(z) = E_{ref}e^{-\gamma z}\hat{\boldsymbol{y}}$, and $\boldsymbol{H}_{ref}(z) = H_{ref}e^{-\gamma z}\hat{\boldsymbol{x}}$, where $\gamma = \omega\sqrt{\varepsilon\mu}$ is the propagation constant.

For the reference PPW, the time-average stored electrical energy over a section of length s along the propagation axis can be expressed as shown in (4.13).

$$W_e^{ref} = \frac{s}{4}\int\int_{\Omega}\varepsilon|E_{ref}(y,z)|^2 dS \tag{4.13}$$

The considered slow-wave PPW is illustrated in Fig. 4.5(b). It is based on the same geometry as the reference waveguide. However, it is assumed that metallic and/or dielectric inclusions are made between the conducting plates in such a way as to confine both electric and magnetic fields in the cross-section sub-domains Ω_e and Ω_m, respectively. In addition, it is assumed that apart from this reduction in occupied space, the field lines shapes are not modified, the complex field vectors $\boldsymbol{E}_{sw}$ and $\boldsymbol{H}_{sw}$ in the slow-wave waveguide are therefore written in equations (4.14) and (4.15) as scaled version of the reference fields.

$$\boldsymbol{E}_{sw}(x,y,z) = \boldsymbol{E}(y,z) = a_e\boldsymbol{E}_{ref}(z)\mathbb{1}_e(y) \tag{4.14}$$

$$\boldsymbol{H}_{sw}(x,y,z) = \boldsymbol{H}(y,z) = a_m\boldsymbol{H}_{ref}(z)\mathbb{1}_m(y) \tag{4.15}$$

where $\mathbb{1}_{e/m}(x,y) = 1$ in $\Omega_{e/m}$ and 0 elsewhere, and a_e and a_m are real constants. These inclusions may be of various shapes and materials, therefore, for simplicity, they are not represented here. Now, let us apply the energy considerations given above to this case. First, the time-average stored electrical energy over a section of length s along the propagation axis can be expressed as shown in (4.16).

$$W_e^{sw} = \frac{s}{4}\int\int_{\Omega}\varepsilon|E_{sw}(y,z)|^2 dS = |a_e|^2\frac{s}{4}\int\int_{\Omega_e}\varepsilon|E_{ref}(z)|^2 dS \tag{4.16}$$

This expression can be further simplified considering that the energy is homogeneously distributed in the reference waveguide transverse section. The surface of the transverse domains Ω, Ω_e and Ω_m, respectively, written as S, S_e and S_m, are introduced, leading to (4.17).

$$W_e^{sw} = |a_e|^2 \frac{sS_e}{4S} \iint_\Omega \varepsilon |E_{ref}(z)|^2 dS = |a_e|^2 \frac{S_e}{S} W_e^{ref} \tag{4.17}$$

Similarly, the derivation can be done for the magnetic components to obtain (4.18).

$$W_m^{sw} = |a_m|^2 \frac{S_m}{S} W_m^{ref} \tag{4.18}$$

Using (4.17), the slow-wave factor (SWF) can be expressed in (4.19).

$$SWF \approx |a_e|^2 \frac{S_e}{S} = |a_m|^2 \frac{S_m}{S} \tag{4.19}$$

Then, the scaling factors ratio can be related to the involved surfaces (4.20).

$$\left| \frac{a_e}{a_m} \right| \approx \sqrt{\frac{Sm}{Se}} \tag{4.20}$$

A second relation can be derived from the fact that both waveguides transmit the same amount of power, as shown in (4.21).

$$\iint_\Omega E_{sw} \times H_{sw}^* \cdot dS = \iint_\Omega E_{ref} \times H_{ref}^* \cdot dS = a_e a_m \iint_{\Omega_e \cap \Omega_m} E_{ref} \times H_{ref}^* \cdot dS \tag{4.21}$$

Because the power is uniformly distributed in the transverse section of the reference waveguide, the term on the right-hand side of (4.21) is a fraction of the total power. Thus, by representing S_{em} as the area corresponding to $\Omega_e \cap \Omega_m$, one obtains equation (4.22).

$$\iint_\Omega E_{sw} \times H_{sw}^* \cdot dS = a_e a_m \frac{S_{em}}{S} \iint_\Omega E_1 \times H_1^* \cdot dS \tag{4.22}$$

This leads to equation (4.23).

$$a_e a_m = \frac{S}{S_{em}} \tag{4.23}$$

Finally, combining (4.20) and (4.23) results in (4.24) and (4.25).

$$a_e = \left(\frac{S}{S_{em}} \right)^{\frac{1}{2}} \left(\frac{S_m}{S_e} \right)^{\frac{1}{4}} \quad a_m = \left(\frac{S}{S_{em}} \right)^{\frac{1}{2}} \left(\frac{S_e}{S_m} \right)^{\frac{1}{4}} \tag{4.24}$$

$$SWF = \frac{\sqrt{S_m S_e}}{S_{em}} \tag{4.25}$$

The magnetic field is much harder to confine than the electric field without magnetic materials, which are often lossy. Therefore, in most practical cases, by confining the electric field, it is possible to completely separate the fields and to control the volume occupied by the intersection of both fields. Hence S_{em} can, in fact, be considered as S_e volume only. Hence, equation (4.25) can be simplified to (4.26).

$$SWF = \sqrt{\frac{S_m}{S_e}} \tag{4.26}$$

The ratio of the surfaces, S_m/S_e, for the same waveguide width, can be expressed as a ratio of thicknesses as presented in (4.27).

$$SWF = \sqrt{\frac{h}{h_e}} \tag{4.27}$$

where h is the total thickness of the waveguide, h_e is the thickness of the region where the electric field is confined.

The expression (4.27) can be used to calculate the effective dielectric constant (4.28) resulting from the electric field confinement, which in turn can be applied to calculate the SIW cut-off frequency (4.29) and the phase velocity (4.30) for the slow-wave SIW.

$$\varepsilon_{r_{eff}} = \varepsilon_r \cdot SWF^2 \tag{4.28}$$

$$f_{c\,SW-SIW} = \frac{c_0}{2 \cdot W_{eff} \cdot \sqrt{\varepsilon_{r_{eff}}}} = \sqrt{\frac{\varepsilon_r}{\varepsilon_{r_{eff}}}} f_{c\_SIW} \tag{4.29}$$

$$v_{\varphi\,SW-SIW} = \frac{c_0}{\sqrt{\varepsilon_{r\_eff} \cdot \left(1 - \left(\frac{f_{c\_SW-SIW}}{f}\right)^2\right)}}. \tag{4.30}$$

These equations offer an approximation resulted from the confinement of electric energy within the transverse section of a SIW, given a specific *SWF*. While these findings are derived in an idealized context, they can offer valuable physical insights applicable to a range of existing slow-wave structures. Comparable expressions can be found in the literature for structures that only partially adhere to the restrictive assumptions employed here.

From an electrical point of view, since the electrical field is confined above the metallic posts, by analogy to a loaded transmission line, it can be considered that the SW-SIW is loaded by a distributed capacitance, which indeed reduces the wave velocity.

Figure 4.6 compares the model from (4.30) with EM simulations for the normalized SW-SIW phase velocity (v_φ/c_0) versus frequency. From these equations, we can see that the cut-off frequency of the SW-SIW decreases proportionally to the square root of thickness' ratio h_e/h. As it can be seen form the figure, the proposed model gives us a first estimate of the waveguide phase velocity and cut-off frequency, even if a non-negligible shift between the calculated cut-off frequencies and the ones obtained by EM simulated exists. Also, the total heigh h and width W_{SIW} were kept constant. A reduction of 40% in $f_{c_{SW-SIW}}$ is obtained by filling 80% of the waveguide with metallic posts. Hence, a narrower SW-SIW could be used for the same targeted cut-off frequency instead of a conventional SIW.

Because the electric field confinement is not ideal, the geometrical aspects of the metallic posts influence both $f_{c_{SW-SIW}}$ and $v_{\varphi\,SW-SIW}$. For instance, the cut-off frequency is inversely proportional to the metallic post diameter as long as the distance between two posts is larger than the confined region, h_e. As this distance reduces, for large post diameters, the magnetic field is increasingly disturbed. Hence, the total magnetic flux decreases when the post diameter increases. From a circuit point-of-view, and in comparison with microstrip lines or Coplanar waveguides (CPWs), this is similar to a decrease of the inductance of the transmission line, leading to an increase of the phase velocity.

Figure 4.6 Influence of the relative height (h_1/h, where $h_1 = h - h_e$) of the metallic posts on the cut-of frequency in PCB technology (Niembro-Martin et al., 2014). Dot line: model simulations; solid line: ANSYS HFSS simulations. Source: Niembro-Martin et al., (2014)/with permission of IEEE.

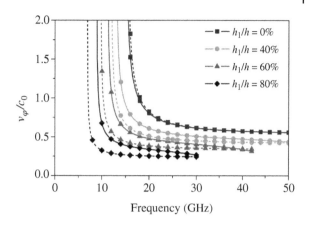

4.3.2 Lossy Slow-Wave PPW (Dielectric Losses)

In the case of low-loss structures, it is presumed that the theorems discussed earlier concerning energy considerations in periodic structures remain applicable. Naturally, as dissipation increases in both the metal and dielectric volumes, the accuracy of these predictions diminishes. We operate under the assumption that the introduction of lossy materials does not substantially alter the field amplitudes, allowing us to obtain a reliable estimation of attenuation based on the lossless field values.

Consider first that both reference (classical) and slow-wave waveguides are carried out with the same lossy dielectric material, with a complex permittivity $\varepsilon = \varepsilon'(1 - j \tan \delta)$. Considering also that the energy separation is performed by metallic posts in the slow-wave structure and that the *SWF* is the ratio between the electrical (or magnetic) stored energy in the slow-wave waveguide and the reference waveguide. Then, it is possible to obtain a first expression of *SWF* (4.31).

$$SWF = \frac{d\iint_{\Omega} \frac{1}{4}\varepsilon' \left\| \boldsymbol{E_{sw}} \right\|^2 dS}{d\iint_{\Omega} \frac{1}{4}\varepsilon' \left\| \boldsymbol{E_{ref}} \right\|^2 dS} = \frac{\iint_{\Omega} \frac{1}{2}\varepsilon'\omega \tan \delta \left\| \boldsymbol{E_{sw}} \right\|^2 dS}{\iint_{\Omega} \frac{1}{2}\varepsilon'\omega \tan \delta \left\| \boldsymbol{E_{ref}} \right\|^2 dS} \tag{4.31}$$

Dividing the numerator and denominator by two times the transmitted power P_t, it is possible to show that the *SWF* corresponds to the ratio of dielectric attenuations, as expressed in (4.32).

$$SWF = \frac{\iint_{\Omega} \frac{1}{2}\varepsilon'\omega \tan \delta \left\| \boldsymbol{E_{sw}} \right\|^2 dS/2P_t}{\iint_{\Omega} \frac{1}{2}\varepsilon'\omega \tan \delta \left\| \boldsymbol{E_{ref}} \right\|^2 dS/2P_t} = \frac{\alpha_{d-sw}}{\alpha_{d-ref}} \tag{4.32}$$

This result highlights the fact that the dielectric attenuation of a slow-wave waveguide is proportional to velocity reduction.

The *SWF* from equation (4.27) was compared with EM simulations for different thickness ratios to validate the modeling in the low dispersion regime, as shown in Figs. 4.7 and 4.8. In particular, Fig. 4.7 allows the validation of equation (4.32), for different height ratios at a given frequency. Indeed, the *SWF* and dielectric attenuation ratio $\alpha(h_{nw})/\alpha(0)$ are similar with a deviation lower than 10% up to a *SWF* of 2.5. Here, h_{nw} is the region

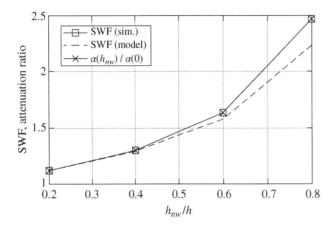

Figure 4.7 SWF and attenuation ratio versus of the thickness ratio at 10 GHz. Black markers represent the SWF calculated from (4.27).

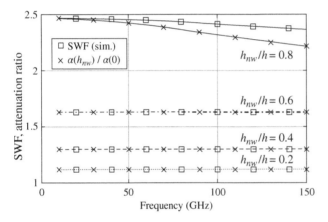

Figure 4.8 SWF and attenuation ratio versus the frequency for different thickness ratios.

with metallic posts (nanowires in the following section) and is equal to the total thickness h minus the region where the electric field is confined, h_e. Besides, Fig. 4.8 illustrates the same frequency behavior for *SWF* and for the attenuation ratio. A very close agreement is obtained between the simulated and theoretically predicted velocity reduction and dielectric attenuation increase when h_{nw}/h is smaller than 0.6.

For a higher thickness ratio (0.8, in the proposed plot), and above 50 GHz, a non-negligible deviation is observed, due to the dispersive nature of the slow-wave propagation, which is not taken into account in the proposed model. Since the dispersion is related to the specific physical and geometrical properties of each slow-wave structure, it can have diverse origins. For instance, in PCB technology, due to non-negligible dimensions of via-holes, the periodic structures obtained for the slow-wave topologies exhibit electrical lengths leading to Bragg effects at few GHz.

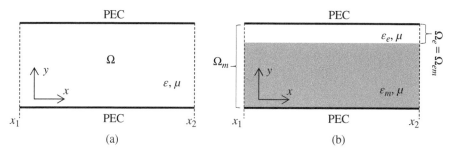

Figure 4.9 Transverse section of both (a) reference and (b) simplified slow-wave PPWs for derivation of the SWF considering metallic losses. A PEC is used in both cases for the parallel plates.

4.3.3 Lossy Slow-Wave PPW (Metallic Posts Losses)

Here, the impact of the vertical conductivity σ_z of the metallic posts on the propagation constant of the SW-SIW is quantified. For simplicity, and based on (Bertrand et al., 2020), a model based on a PPW is proposed and validated by EM simulations. The top and bottom metallic surfaces of the PPW are considered lossless. In practical cases, losses in vias (metallic posts) are always significantly higher.

This theoretical advancement is rooted in the concept that the confinement of the electric field is less than perfect, and it relies on the electrical characteristics of the dielectric materials that make up the slow-wave waveguide.

The slow-wave PPW illustrated in Fig. 4.5 is simplified by considering a transverse domain constituted only by two sub-domains Ω_e and Ω_m, with corresponding surfaces S_e and S_m, respectively. Here, S_{em} is equal to $S_{e'}$, as shown in Fig. 4.9.

A TEM propagation is considered along the x-axis. The same lossless dielectric material as the PPW one (see Fig. 4.9) is considered in the upper part of the SW-PPW (of cross-section S_e and thickness h_e) with a relative dielectric constant $\varepsilon_{re} = \varepsilon_r$, where a dielectric material with losses is used in the bottom part (of cross-section S_m and thickness h_m) with a complex relative permittivity $\overline{\varepsilon_{rm}}$ as defined in (4.33). The real part ε'_{r_m} corresponds to the dielectric constant of the material and the imaginary part ε''_{r_m} defined in (4.34) allows taking losses into account by considering the vertical conductivity σ_z, angular frequency ω, and loss tangent $\tan\delta$ for the dielectric dissipation of this material, with ε_0 being the vacuum dielectric constant.

$$\overline{\varepsilon}_{rm} = \varepsilon'_{r_m} - j\,\varepsilon''_{r_m} \tag{4.33}$$

$$\varepsilon''_{r_m} = \frac{\sigma_z}{\omega\varepsilon_0} + \varepsilon'_{r_m} \cdot \tan\delta \tag{4.34}$$

In this context, the time-averaged stored electrical and magnetic energies are treated as complex numbers for the slow-wave structure to account for power loss, as described by (Collin, 1990), resulted from the finite equivalent vertical conductivity (σ_z) of the metallic posts. Consequently, we introduce a complex slow-wave factor ($\overline{SWF}$) in the following equation (4.35).

$$\overline{SWF} = \frac{\overline{W}_e}{W_e^{ref}} = \frac{\overline{W}_m}{W_m^{ref}} \tag{4.35}$$

The complex slow-wave factor $\overline{SWF}$ is related to an equivalent relative effective complex dielectric constant $\overline{\varepsilon_{r_{eff}}}$ of the slow-wave medium of interest, defined in (4.36). This complex dielectric constant is composed of a real effective dielectric constant $\varepsilon_{r_{eff}}$ expressed in (4.37) and an effective loss tangent $\tan\delta_{eff}$ defined in (4.38), which simulates losses in the structure.

$$\overline{\varepsilon_{r_{eff}}} = \overline{SWF}^2 = \varepsilon_{r_{eff}} \cdot (1 - j\tan\delta_{eff}) \tag{4.36}$$

$$\varepsilon_{r_{eff}} = Re(\overline{SWF}^2) \tag{4.37}$$

$$\tan\delta_{eff} = -\frac{Im(\overline{SWF}^2)}{\varepsilon_{r_{eff}}} \tag{4.38}$$

The $\overline{SWF}$ for the SW-PPW illustrated in Fig. 4.11 is proposed in (4.39), considering the assumptions and definitions detailed above.

$$\overline{SWF} = \sqrt{\frac{S_e + S_m}{S_e + S_m\frac{\varepsilon_{r_e}}{\overline{\varepsilon_{r_m}}}}} \tag{4.39}$$

Finally, equation (4.40) allows considering only the thicknesses and dielectric properties. When considering a high vertical conductivity, leading to ε''_{r_m} much greater than ε'_{r_m}, the complex $\overline{SWF}$ tends to the real SWF of (4.27).

$$\overline{SWF} = \sqrt{\frac{h_e + h_m}{h_e + \frac{h_m \cdot \varepsilon_{r_e}}{\varepsilon'_{r_m} - j\varepsilon''_{r_m}}}} \tag{4.40}$$

In order to validate the expressions of the relative effective dielectric constant $\varepsilon_{r_{eff}}$ in (4.37) and effective loss tangent $\tan\delta_{eff}$ (4.38), the MnM technology described in Chapter 3 (Section 3.4) was considered. Let us notice that the copper nanowire conductivity (vertical conductivity, σ_z) can strongly deviate from the bulk conductivity for this specific technology. Hence, Eigen-mode type electromagnetic simulations using ANSYS High Frequency Structure Simulation (HFSS) were performed.

Due to both the high density of nanowires in the MnM technology ($>5\cdot10^{13}$ wires/m^2), and the large form factors existing in the structure (nanometric pore diameter d_p and spacing e_p, micrometric dielectric thickness and millimetric lateral dimensions of the waveguide), the development of an equivalent medium to the membrane filled with copper was required. Based on a model proposed for arrays of nanotubes (Franck et al., 2012), an anisotropic but homogeneous model can be derived for the nanowire membrane.

In the proposed model, the membrane is considered as a dielectric allowing perfectly insulating adjacent nanowires from each other. Hence, the membrane can be considered as a dielectric in the x- and y-axis (with relative dielectric constant of 9.7 – considering bulk alumina – and dissipation factor of 0.015 – extracted from measurements).

Along the vertical z-axis, an equivalent conductivity σ_z is considered to take the conduction of the nanowires into account. This conductivity (4.41) is proportional to the copper conductivity σ_{cu} and the membrane porosity δ_p. Based on Scanning Electron Microscopy (SEM) observations of nanoporous membranes (Franc et al., 2012), an hexagonal regular distribution is observed, as drawn in Fig. 4.10 with a porosity δ_p defined as the ratio of the

Figure 4.10 Illustration of the microscopic top view of the nanoporous membrane filled with copper.

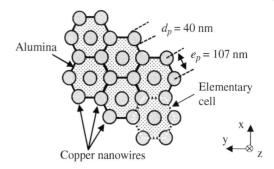

pore area A_{nw} per elementary cell (4.42) over the elementary cell area A_{cell} (4.43), depending on both pore diameter d_p and on pore spacing e_p. Considering the physical properties of the nanowire membrane, the estimated equivalent vertical conductivity can reach 3.8 MS/m. The accuracy of this simulation approach was established through a comparison with the measurement data derived from the slow-wave microstrip lines from (Serrano et al., 2014).

$$\sigma_z = \frac{A_{nw}}{A_{cell}} \cdot \sigma_{cu} = \delta_p \cdot \sigma_{cu} \tag{4.41}$$

$$A_{nw} = 3\pi \left(\frac{d_p}{2}\right)^2 \tag{4.42}$$

$$A_{cell} = \frac{3\sqrt{3}}{2} e_p^2 \tag{4.43}$$

As presented by (Corsi et al., 2020), the impact of the equivalent vertical membrane conductivity σ_z on the relative effective dielectric constant $\varepsilon_{r_{eff}}$ can be observed for different values of the air layer thickness $h_e = h_{air}$, from 5 to 20 μm, as shown in Fig. 4.11. A membrane thickness h_{nw} of 50 μm with nanowires acting as metallic posts is considered. The behavior of $\varepsilon_{r_{eff}}$ obtained by the proposed model (4.37) and the one obtained from EM simulations are in good agreement for vertical conductivities varying from $1 \cdot 10^{-3}$ kS/m to $1 \cdot 10^3$ kS/m. Besides, Fig. 4.11 allows highlighting that, whatever the value of h_{air}, the relative effective dielectric constant presents two levels (the one at low σ_z and the second one at high σ_z) separated by a transition zone located close to 1 kS/m. Let us notice that the value of $\varepsilon_{r_{eff}}$ at high σ_z corresponds to the value given by (4.28) without taken σ_z into consideration.

Figure 4.11 shows also the electric and magnetic fields in a lateral cross-sectional view of the SW-PPW for two conductivities, 1 kS/m and $1 \cdot 10^3$ kS/m, respectively. When a high conductivity ($1 \cdot 10^3$ kS/m) is considered, there is a strong confinement of the electric filed in the air layer and no propagation in the bottom dielectric. Whereas, the magnetic field is almost not perturbed: hence a slow-wave mode is achieved. On the other hand, when the conductivity is small, 1 S/m (Fig. 4.11), the electrical field propagates in both air and dielectric layers.

At the interface of the membrane and air the magnitude of the electric field is lower in the alumina membrane, since its dielectric constant is equal to 9.7. Consequently, the separation between the electric and magnetic fields is no longer maintained, resulting in the loss of the structure's slow-wave characteristics. Therefore, the $\varepsilon_{r_{eff}}$ is higher in the high conductivity condition, than in the lower conductivity one.

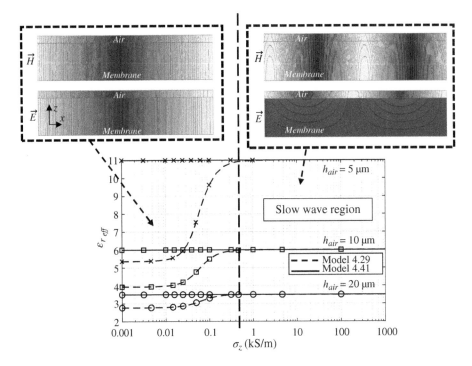

Figure 4.11 Slow-wave PPW: effect of the equivalent vertical membrane conductivity σ_z on the relative effective dielectric constant $\varepsilon_{r_{eff}}$. Relative effective dielectric constant $\varepsilon_{r_{eff}}$ as a function of σ_z for different air thicknesses $h_e = h_{air}$ (with $h_{nw} = 50\ \mu m$, $\varepsilon'_{r_{m2}} = 9.7$, $\tan \delta = 0.015$).

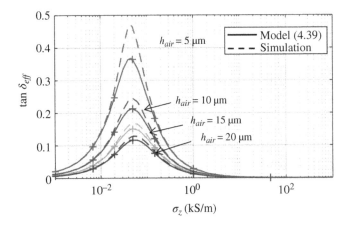

Figure 4.12 Slow-wave PPW: effect of the equivalent vertical membrane conductivity σ_z on the effective dissipation factor $\tan \delta_{eff}$ for different air thicknesses $h_e = h_{air}$ (with $h_{nw} = 50\ \mu m$, $\varepsilon'_{r_{m2}} = 9.7$, $\tan \delta = 0.015$).

For high values of conductivity, not taking into consideration σ_z, the calculated values of $\varepsilon_{r_{eff}}$ using equation (4.37) are similar to those obtained in EM simulations. This means, as shown in Fig. 4.11, that if the conductivity is higher that 1 kS/m, the $\varepsilon_{r_{eff}}$ can be approximated using equation (4.28), since the slow-wave propagation mode is maintained.

Figure 4.12 illustrates the impact of σ_z on the effective loss tangent $\tan\delta_{eff}$ calculated using (4.38). The observed behavior is compared with EM simulation results for different air layer thicknesses h_{air}, from 5 to 20 µm. A good agreement is obtained, thus validating the proposed model.

High values of σ_z (higher than 10^1 kS/m) lead to low metallic losses of the nanowires and strong confinement of the electric field in the air layer, which, in turn, are reflected in a low $\tan\delta_{eff}$. Also, very low vertical conductivity of the nanowires (below 10^{-3} kS/m) is related to a low $\tan\delta_{eff}$, since the porous alumina filled with nanowires behaves entirely as a dielectric.

As shown in Fig. 4.12, the $\tan\delta_{eff}$ reaches a peak at approximately $5\cdot10^{-2}$ kS/m of σ_z, independently of the thickness of h_{air}. This peak value of $\tan\delta_{eff}$ occurs at the transition from a classic SIW propagation mode to a slow-wave propagation mode, as illustrated in Fig. 4.11.

4.4 SW-SIW in PCB Technology

4.4.1 Design Rules

Figure 4.13 illustrates a cut 3D-view of standard PCB technology with three metallic layers for the fabrication of SW-SIW. Based on parametric EM simulations, design rules can be defined depending on a miniaturization goal, i.e. a given value of SWF. In order to allow an easy comparison between theory and simulations, consider a technology stackup which does not contain dielectric constant variations. In circuit design, one should keep in mind that the occupied area of a SW-SIW is about that of its counterpart in SIW divided by SWF^2, since the miniaturization is both in transversal and longitudinal dimensions.

First of all, considering the standard copper conductivity and dielectric constants available, a reasonable upper limit for SWF can be set at two, which leads to a surface reduction by four. For higher SWF, the degradation of the quality factor is important, since the quality factor decreases with the SWF, as shown in Section 4.3. Once the SWF value is defined, an estimation of the maximum spacing S_1 between blind vias can be obtained by aiming at a distributed effect for easier circuit design. Considering an operation frequency f placed

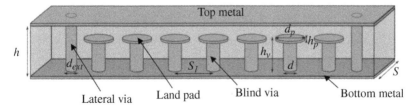

Figure 4.13 Cut view of standard PCB technology with three metallic layers for the fabrication of SW-SIW.

at the center of the mono-mode frequency band, the cut-off frequency of the TE_{10} mode is therefore defined as $fc = 2/3f$. The maximum frequency in the mono-mode operation is $2fc$, it is associated to the smaller guided wavelength $\lambda(2fc)$. For instance, a distributed assumption can be expressed by $\lambda(2fc) \leq 10S_1$. Furthermore, based on the expression of β given by (4.9), one can easily derive equation (4.44).

$$\lambda(2f_c) = \frac{c_0/\sqrt{3\varepsilon_r}}{f_c \cdot SWF} \tag{4.44}$$

Then, the following upper limit for S_1 can be defined:

$$S_1 \leq \frac{c_0}{10 \cdot \sqrt{3\varepsilon_r} \cdot f_c \cdot SWF} \tag{4.45}$$

The lower limit for S_1 was not defined explicitly, however reducing S_1 quickly leads to an increased attenuation and degraded quality factor. The upper limit as defined by (4.45) can, for instance, be adopted as initial value.

The same spacing S_1 can be used for the distance between vias in the lateral walls for simplicity of design and simulation, even if the upper limit fixed for S_1 is more restrictive than that given by (4.6). The definition of the lateral walls can be completed by setting the diameter of the external vias (d_{ext}) to $S_1/2$, as it is often chosen in SIW. Finally, using the expression of the effective width (4.10) and the cut-off frequency value, the waveguide lateral dimension W can be estimated.

In general, metallic layer thickness follows standard format, so that the choice of the thickness of the land pad h_p is already pre-defined by the PCB manufacturer. Then, h_v and h can be defined also considering the standard commercialized substrate thicknesses. As a first approximation, they can be related to the SWF by the following expressions (4.46) and (4.47), which is readily derived from (4.26) by neglecting the fringing field.

$$SWF = \sqrt{\frac{h}{h - (h_v + h_p)}} \tag{4.46}$$

$$\varepsilon_{r_{eff\,SW}} = \varepsilon_{r_{eff\,SIW}} \left(\frac{h}{h - (h_v + h_p)} \right) \tag{4.47}$$

Depending on the material availability, a combination of h and h_v leading to the proper SWF can be defined.

In the end, the only parameters remaining are the diameters d and dp. Already, fabrication constraints generally impose that dp should be greater than d with a certain margin in order to ensure a functional copper plating process. The rule $dp \geq d + 200\,\mu m$ is a good starting point. Based on the parametric analysis, the following rules can be applied. First, for a given SWF, thinner via holes do provide greater quality factors, at the condition of reducing the copper land diameter accordingly. This procedure should be finalized by performing a simulation of the obtained waveguide, to confirm and if necessary, adjust the dimensions.

4.4.2 Ku-Band SW-SIW Implementation and Results

In (Niembro-Martin et al., 2014), a first SW-SIW was designed, manufactured and measured to validate the slow-wave concept in SIW technology. For that, two layers of Rogers substrates, *Sub1* and *Sub2* (RO4003 and RO4403) with thicknesses h_1 = 0.813 mm and h_2 = 0.3 mm, with relative dielectric constant ε_{r1} = 3.55 and ε_{r2} = 3.17, dielectric loss tangent $\tan \delta_1$ = 0.0027 and $\tan \delta_2$ = 0.005, respectively, and both with a copper thickness of 50 µm were used. Limited by the PCB manufacture capability and guided by numerical simulations, d = 400 µm, S = 800 µm, and S_1 = 900 µm were used.

To obtain initial values for a 3D EM simulation, equations (4.29) and (4.30) were used. By defining a cut-off frequency of 8.4 GHz and considering the dielectric constant as the one from Sub_2, since almost all the electric field is confined in this substrate, the lateral dimension W_{SIW} of the SW-SIW of 5.2 mm is obtained. For this width and the technological constrains for the via fabrication, five rows of blind vias were used in the design. Further, for manufacturing purposes, land pads with diameter of 600 µm, were added on the top of Sub_1 (*Metal 2*), as shown in Fig. 4.14(a).

Microstrip lines with taper sections were added to feed the waveguide, as shown in Fig. 4.14(b).

To concentrate the main part of the electric field in Sub_2, internal blind vias were also used under the 50-Ω microstrip line tapered sections. This makes it easier to transfer the electromagnetic wave into the SW-SIW. The microstrip line width W_s of 1 mm resulted in a 50-Ω impedance. Then, the width W_{tap} and length L_{tap} of the tapered sections were optimized by EM simulations to improve the return loss in the largest bandwidth (W_{tap} = 5.5 mm and L_{tap} = 2.4 mm).

Figure 4.15 shows the improvement in matching of the SW-SIW when using the optimized tapered sections. Indeed, with the tapered sections, a return loss better than 10 dB is achievable in the mono-mode bandwidth between 12 and 20 GHz, as presented in Fig. 4.15(b). The fabricated 16-mm-long SW-SIW is shown in the photograph of Fig. 4.16.

Measurement results are shown in Fig. 4.17. A good agreement between S-parameter measurements and the EM simulations in ANSYSY HFSS was obtained. The measured cut-off frequency is 9.3 GHz, which corresponds to a frequency shift of 15% with the predicted cut-off frequency calculated using (4.29). This measured cut-off frequency is 43% smaller than the classical SIW one (f_{c_SIW} = 16.2 GHz) designed with the same lateral width W_{SIW}, thus validating the high transversal miniaturization of the SW-SIW. Thanks to the optimized tapered sections, a return loss better than −12.7 dB is obtained from

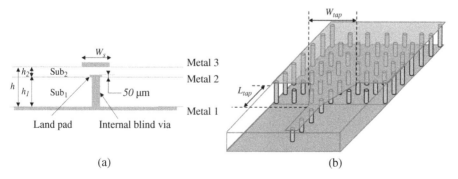

(a) (b)

Figure 4.14 Schematic view of the proposed microstrip/SW-SIW tapers: (a) cross-section view of the microstrip feeding lines and (b) 3–dimensional view.

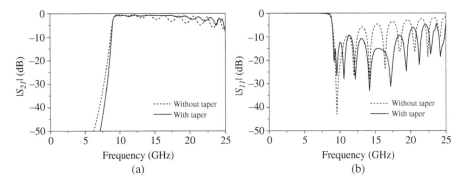

Figure 4.15 Simulated S-parameters of the 16-mm long SW-SIW with and without tapering sections. Source: Niembro-Martin et al., (2014)/with permission of IEEE.

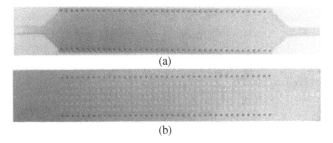

Figure 4.16 Photograph of the fabricated SW-SIW with tapers: (a) Top view and (b) bottom view. Source: Niembro-Martin et al., (2014)/with permission of IEEE.

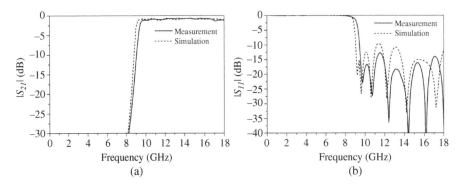

Figure 4.17 S-parameters of the 16-mm long SW-SIW. Comparison between measurement and simulation results. (a) insertion loss and (b) return loss.

9.5 to 18 GHz. Finally, the total insertion loss (including the tapers' one) equals 0.63 dB at 14.6 GHz.

Based on the above S-parameters, and by using the method proposed in (Mangan et al., 2006), the attenuation constant, phase constant and phase velocity of the SW-SIW can be extracted. Again, a good agreement can be observed between measurements and EM simulation, as shown in Fig. 4.18. The peaks in the attenuation constant in Fig. 4.18(a) are

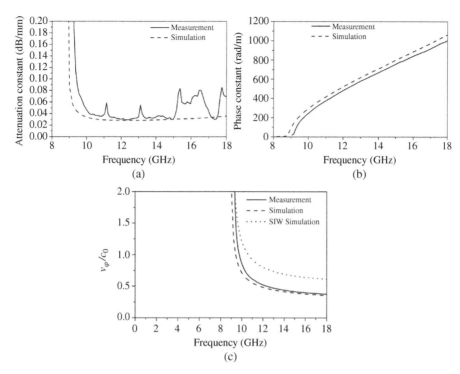

Figure 4.18 (a) Attenuation constant, (b) phase constant versus frequency for the SW-SIW and (c) normalized phase velocity. Measurement and simulation results.

related to the resonances of the electrical length of SW-SIW at multiple of the half-guided wavelength. The SIW and SW-SIW phase velocity are compared in Fig. 4.18(a), (b) and (c): the phase velocity of the SW-SIW is about 40% smaller than its classical SIW counterpart.

Finally, when comparing the lateral dimensions of both SW-SIW and SIW designed for the same cut-off frequency, the SW-SIW width is about 40% smaller than the SIW one, and finally, the surface area reduction is greater than 60%.

4.4.3 SW-SIW Coupler

The slow-wave topology, which was introduced in the previous sections, can be used to design 3-dB couplers. Such a coupler was designed by (Bertrand et al., 2019b) in Ku-Band (11 GHz). The design methodology and measurement results are described in this section.

A ratio $\eta = h_2/(h_2 + h_1)$ of 0.3 was chosen to achieve a reasonable compromise between *SWF* and insertion loss, which resulted in an effective dielectric constant approximately three times higher than the intrinsic dielectric constant of the material (Niembro-Martin et al., 2014).

For *Sub1* with blind vias, a Rogers RO4003 substrate ($\varepsilon_r = 3.55$, $\tan \delta = 2.710^{-3}$) with $h_1 = 0.813$ mm was chosen. For Sub2, a stack of two materials (RO4003 and RO4450 pre-peg – $\varepsilon_r = 3.54$, $\tan \delta = 4.10^{-3}$) with $h_2 = 0.203 + 0.102$ mm was used. RO4450 is used to attach the top RO4003 without vias to the bottom one with vias.

The center to center distance in the transverse and longitudinal directions between blind vias was defined as 0.8 mm, to create an isotropic slow-wave effect. To comply with fabrication limitations, copper land pad diameter d_p was defined as 0.6 mm with a thickness h_p of 35 μm and a via diameter d of 0.4 mm. To avoid leakage or unwanted periodicity effect because of the guided reduced wavelength, the diameter d_{ext} of the lateral vias was defined as 0.2 mm instead of 0.4 mm.

Figure 4.19 presents a 3D view and a top view of the SW-SIW coupler. By choosing a width W_a of the feeding waveguide equal to 6.2 mm with 7 blind vias in cross section, the cutoff frequency is equal to about 7.5 GHz. In this configuration, at 11.2 GHz the insertion loss is minimum, according to EM simulation using an eigenmode analysis in ANSYS HFSS. This simulation also provides a slow-wave wavenumber k_{sw} centered at 11 GHz. Using equations (4.48) and (4.49), with n being a positive integer, the coupling regions' dimension is calculated to $W = 10.65$ mm and $L = 6.87$ mm ($n = 1$), as presented in (Chen & Chu, 2010). These dimensions were used in EM simulations for optimization of the performance of the SW-SIW coupler.

$$L = \frac{\pi}{k_{sw}} \sqrt{\frac{4(3n+1)(n+1)}{3}} \tag{4.48}$$

$$W = \frac{\pi}{k_{sw}} \sqrt{\frac{(4n+1)(n+1)}{4n+1}} \tag{4.49}$$

In these equations (4.48) and (4.49) the use of k_{sw} is based on the assumption that the phase velocities of the two modes TE_{10} and TE_{20} are reduced by the same $SWF v_\varphi^{sw-siw}/v_\varphi^{siw}$. The reason for this is that the guided wavelength is significantly greater than the separation between the vias, with the separation s in this scenario being approximately one-tenth of a wavelength. Additionally, the transverse section is fully occupied by blind vias, ensuring that the slow-wave effect is preserved at all locations where the electric field is at its maximum.

To reduce EM simulation time in order to optimize the SW-SIW coupler, an equivalent homogeneous substrate was defined based on the uniformly distributed slow-wave effect.

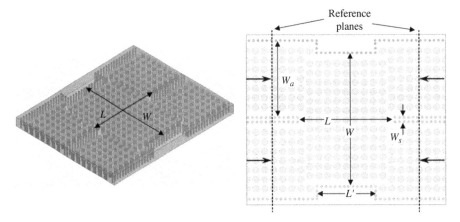

Figure 4.19 Geometrical parameters of the 3-dB coupler in SW-SIW technology. Source: Adapted from Bertrand et al., (2019a).

For this, an increased dielectric constant, where $\varepsilon_{reff} = \varepsilon_r \cdot SWF^2$ was considered. After the coupler's performance optimization using the equivalent substrate, the real structure was simulated and small adjustments were made.

From the initial dimensions calculated using (4.48) and (4.49), W and L were fixed to 11.45 and 8.58 mm, respectively, after optimization. As explained in (Chen & Chu, 2010), this is essential to reduce the effect of the evanescent TE_{30} mode. Finally, the separation between adjacent waveguides W_s and the length of the narrow region L' were fixed to 0.6 and 5.14 mm, respectively.

Broadband transitions as well as feeding lines were implemented for measurements purposes. A similar topology for SIW and SW-SIW couplers was used for comparison. As in the microstrip to SW-SIW feeding line topology presented above, blind vias were placed below signal strip of a Grounded CPW in order to better match the electrical field distribution of the SW-SIW, so that most of the electric field lines are confined in the upper layers, as shown in Fig. 4.20, were the geometric parameters of this configuration are given.

A transition to the SW-SIW was realized by tapering of the central conductor as depicted in Fig. 4.20. In this figure the top view of the transition is represented with the top copper in partial transparency. The initial length of the tapered section L_{taper} was set to a quarter guided wavelength at 11 GHz, which is 3.62 mm. Vias were kept along the slot as for the feeding line, following the tapered shape. Then, full-wave simulations were performed in order to minimize return loss around the operating frequency. The final dimensions are $L_{taper} = 4.18$ mm and $\theta_{taper} = 32$ degrees.

The method described above, used to design the SW-SIW coupler was used in a conventional SIW coupler for comparison. For the latter, $W = 19.7$ mm, $L = 14.92$ mm, $L' = 8.8$ mm,

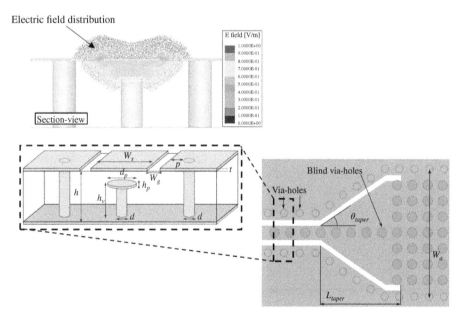

Figure 4.20 Electric field distribution and topology of the loaded Grounded CPW feeding line and transition to SW-SIW.

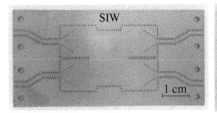 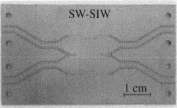

Figure 4.21 Photographs of the SIW (left) and SW-SIW (right) couplers manufactured with RO4003 substrate. Source: Bertrand et al., (2019a)/with permission of IEEE.

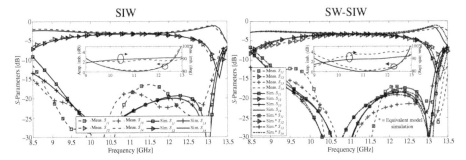

Figure 4.22 Comparison of the Simulated and measured S-parameters of the 3-dB coupler based *on SIW and SW-SIW*. Source: Bertrand et al., (2019a)/with permission of IEEE.

and $W_a = 11.16$ mm were obtained after optimization. Feeding to the couplers was carried out with conventional Grounded CPWs and tapers. Figure 4.21 presents photographs of the fabricated SIW and SW-SIW couplers, respectively.

The reference planes were placed at a quarter-wavelength from the coupling region in order to compare the couplers. A TRL calibration was used in both measurement and simulation. A four-port Anritsu VectorStar MS4647 VNA with end-launch connectors was used for measurements. Figure 4.22 shows a good agreement between simulated and measured performance for the conventional SIW and SW-SIW couplers.

The fabrication process tolerances lead to a small shift in the measured center frequency for the couplers, as it can be in seen in Fig. 4.22. The via's landing pad diameter d_p and thickness h_p were measured to be 0.5 mm and 46 μm, respectively, resulting to the observed frequency shift. The measured insertion loss is 0.4–0.6 dB for the SW-SIW coupler, 0.1 dB for the SIW coupler, respectively. The relative bandwidth is 42 and 43%, respectively, and the return loss is better than 10 dB for both couplers. A measured 1-dB amplitude imbalance bandwidth of 18 and 21% were measured for the SW-SIW and SIW couplers, respectively. The SW-SIW coupler presents a slightly higher dispersion, which can be seen in the bandwidth determination when considering a phase variation of ±2.5°. In this case, the measured bandwidths were 36 and 25% for the SIW and SW-SIW couplers, respectively. Despite the increase dispersion and insertion loss, the surface, normalized to the dielectric wavelength, is reduced by a factor of 3. Like with all slow-wave structures, there is a trade-off between miniaturization and electrical performance, as losses increase with the *SWF*, as demonstrated in Section 4.3 (Bertrand et al., 2020).

4.4.4 SW-SIW Cavity Filter

The slow-wave SIW concept described above was applied to the design of a bandpass cavity filter in the Ku-band (Bertrand et al., 2015). As a first design, the topology of the SIW bandpass filter in (Chen & Wu, 2008) was used. In this proof-of-concept, the study aims to demonstrate that, considering comparable dimensions for both SIW and SW-SIW bandpass filters, the working frequency of the SW-SIW filter is much lower than the SIW one.

A Rogers RO4003 substrate was used as *Sub1* with a thickness of 0.813 mm and RO4403 bonding ply was used as *Sub2* with a thickness of 0.29 mm. The via diameter of 400 μm with center-to-center spacing of 0.8 mm were considered. As previously discussed, in PCB technology, creating internal blind vias necessitates the inclusion of a copper pad on the top of each hole. This additional conductor has a thickness of approximately 50 μm, consequently reducing the separation between the top copper layer and the internal via. The filters were fine-tuned through EM simulations using ANSYS HFSS. This process is required for comparing, on the same substrates, the SW-SIW filter and the conventional SIW filter.

The dimensions of the optimized filters are illustrated in Fig. 4.23. The filter return loss was improved by using slots in the microstrip to SIW transitions.

The designed dimensions of the classical filter are as follows: cavity lengths $L_1 = 5.40$ mm, $L_2 = 5.45$ mm, $L_3 = 5.52$ mm and $L_{slot} = 1.32$ mm. Apertures' sizes are $W_{12} = 2.67$ mm and $W_{23} = 2.47$ mm. The cavity width is $W = 6.07$ mm. These dimensions must be compared to the ones of the SW-SIW optimized filter, which are $L_1 = 4.97$ mm, $L_2 = 5.5$ mm, $L_3 = 5.66$ mm and $L_{slot} = 1.1$ mm. Apertures' sizes are $W_{12} = 2.78$ mm and $W_{23} = 2.50$ mm. The cavity width is $W = 5.7$ mm. Hence the dimensions of both filters are almost similar.

Figure 4.24 presents the EM simulations results for the SIW filter and the EM simulations results and measurement results of the SW-SIW filters. The SIW filter presents a 5.95% relative bandwidth centered at 20 GHz, 2.7 dB insertion loss and better than 18 dB return loss; lossless EM simulations of this filter permit to highlighting the five transmission zeroes. The SW-SIW filter is centered at 11 GHz and presents a 8% relative bandwidth with a 3.2 dB insertion loss. The measured return loss is better than 12 dB. As shown in Fig. 4.23, blind

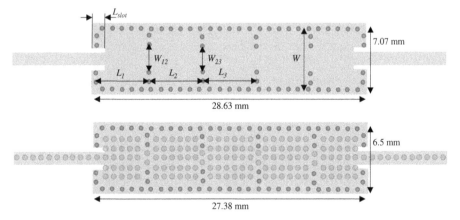

Figure 4.23 Geometrical parameters of the classical SIW cavity filter (Top) and SW-SIW cavity filter (Bottom).

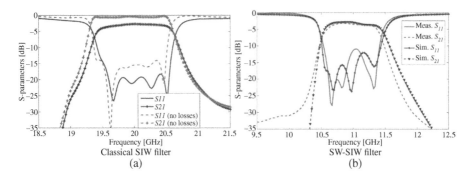

Figure 4.24 Simulated S-parameters of the SIW cavity filter (left) and simulation and measurement of the SW-SIW cavity filter. Source: Bertrand et al., (2015)/with permission of IEEE.

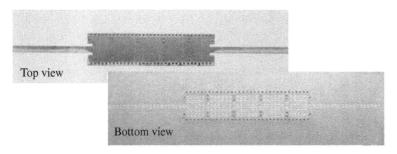

Figure 4.25 Fabricated SW-SIW cavity filter: top and bottom views. Source: Bertrand et al., (2015)/with permission of IEEE.

vias were placed beneath the feeding microstrip lines to concentrate the electric field within *Sub2*, thereby improving the transition to the SW-SIW cavity filter. This step was required to obtain a good impedance matching.

The area of the SW-SIW cavity filter, shown in Fig. 4.25, is $1.7\,\text{cm}^2$, which corresponds to a size reduction of approximately 70% as compared to similar SIW cavity filters.

The design of such a SW-SIW cavity filter is very time consuming, since it is based on full-wave EM simulations. Therefore, a synthesis and optimization method was developed to improve the response of the SW-SIW filter. Details of this method is presented in (Bertrand et al., 2018).

Since it is necessary to perform an iterative optimization to obtain an optimized structure, and since this optimization can turn out to be very time-consuming, especially for high-order filters based on complex structures such as SW-SIWs, a synthesis method based on the segmentation of the filter with cascaded matrices was developed.

Based on this synthesis process, the performance of the filter response can be optimized. The resulting fabricated filter is illustrated in Fig. 4.26. The optimized filter dimensions are $L_1 = 4.34\,\text{mm}$, $L_2 = 5.04\,\text{mm}$, $L_3 = 5.19\,\text{mm}$. Apertures' sizes are $W_{01} = 2.01\,\text{mm}$ $W_{12} = 1.42\,\text{mm}$ and $W_{23} = 1.29\,\text{mm}$.

In this design, Grounded CPW transitions were used to improve matching. Fabrication tolerances resulted in different dimensions of the blind vias. Here, $d_p = 0.5\,\text{mm}$ and $h_p - 46\,\mu\text{m}$, as before, instead of $d_p = 0.6\,\text{mm}$ and $h_p = 35\,\mu\text{m}$. Also, a copper roughness of

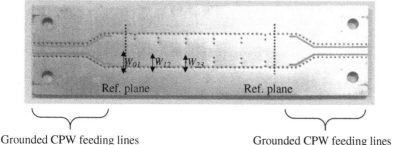

Figure 4.26 Realized filter of a 1-GHz bandwidth in SW-SIW technology. Source: Bertrand et al., (2018)/with permission of IEEE.

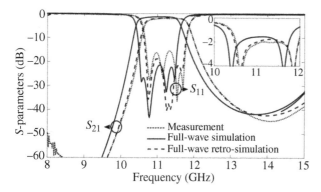

Figure 4.27 Comparison between measurement, simulation, and retro-simulation for a 1 GHz bandwidth filter in SW-SIW technology. Source: Bertrand et al., (2018)/with permission of IEEE.

5 μm was also considered. The EM simulations were adjusted according to the fabricated dimensions and compared with measurements, as shown in Fig. 4.27.

The fabricated 1-GHz bandwidth filter is centered at 11.24 GHz and the measured bandwidth is 1.06 GHz (versus 1.02 GHz in EM simulation). The measured insertion loss is 1.9 dB (versus 1.55 dB in simulation) and the return loss between 10.7 GHz and 11.75 GHz is better than 15.5 dB. In conclusion, it is evident that accounting for process variations in simulations accurately reproduces both the degradation in return loss and the frequency shift. To achieve even better results, a higher level of process control is recommended to ensure greater precision.

4.4.5 Slow-Wave SIW Cavity-Backed Antenna

The concept of SW-SIW can also be used to successfully design compact antennas. In (Ho et al., 2021), a SW-SIW cavity-backed circular-polarized antenna was realized in PCB technology using a slow-wave resonant structure to reduce antenna size. This slow-wave cavity backed square slot antenna (SW-CBSA) was designed by using an inductive via hole asymmetrically inserted into the patch surface, as proposed in (Kim et al., 2011), to obtain a circular polarization and increase the impedance bandwidth.

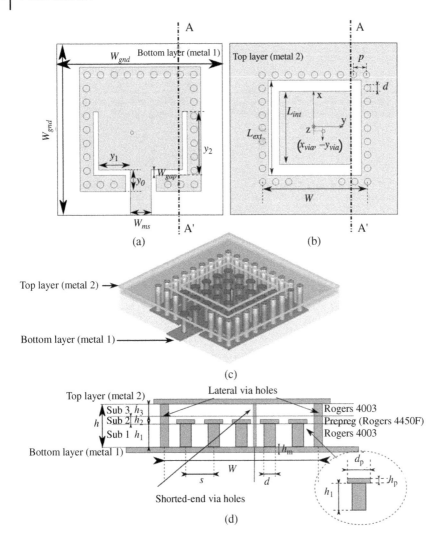

Figure 4.28 Geometrical description of the SW-SIW antenna. (a) Bottom view, (b) Top view, (c) 3-D view and (d) Cross-sectional view. Source: Ho et al., (2021)/with permission of IEEE.

The considered antenna topology is shown in Fig. 4.28. A square ring slot is etched in the top Metal 2 layer of the SIW cavity to excite the first mode TE_{100}. Here again, the slow-wave effect is obtained by inserting internal blind vias connected to the bottom Metal 1 layer, as shown in Fig. 4.28(c) and (d). For the fabrication, two substrate layers of Rogers RO4003 (relative dielectric constant of 3.55, loss tangent of 0.0027, thicknesses $h_1 = 0.813$ mm, and $h_3 = 0.203$ mm, respectively), an adhesive layer, Rogers 4450F (relative dielectric constant of 3.52, loss tangent of 0.004, and thickness $h_2 = 0.102$ mm, respectively) were being used to attached the two RO4003 substrates. The top and bottom metallic layers have a thickness $h_m = 50$ µm, as well as the intermediate metallic layer h_p.

Figure 4.29 presents the impact of the number of blind vias on the miniaturization, the quality factor, and losses (dielectric and metallic ones). These EM simulations were carried

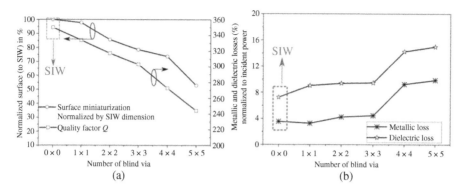

Figure 4.29 (a) Miniaturization (normalized by SIW cavity dimension) and quality factor of the SW-SIW cavity and (b) Metallic and dielectric losses versus the number of the blind vias. Source: Ho et al., (2021)/with permission of IEEE.

out using CST Studio and highlight that the greater the number of blind vias, the smaller the surface of the antenna. This is consistent with previous studies, since a higher confinement of the electric field in the top volume of the SW-SIW structure is achieved, leading to an increased slow-wave effect. However, this miniaturization is accompanied by an increase in losses, which is consistent with the equation (4.32), when considering dielectric losses in the substrates. Hence, a high slow-wave effect leads to both smaller size and higher losses. Figure 4.29(b) underlines the strong impact of the number of blind vias on the dielectric and metallic losses, respectively. As for all slow-wave structures, as underlined in (Bertrand et al., 2020), a compromise between compactness and total efficiency.

In the fabricated antenna, an internal via array of 5×5 was chosen, leading to a surface area reduction of 53% as compared to the SIW counterpart). As explained before, a lower quality factor ($Q = 242$) is achieved as compared to the SIW one ($Q = 350$).

Based on (Janaswamy & Schaubert, 1986), the operating frequency f_{0_SW} of the SW-SIW cavity can be determined as (4.50) in which $\varepsilon_{r_{eff}\_SW}$ is the effective dielectric constant given in (4.47) and $\varepsilon_{r_{eff}}$ corresponds to the effective relative dielectric constant of the dielectric layers between the top of the land pads and the Metal 2 layer.

$$f_{0\_SW} = \frac{c_0}{2(L_{int} + L_{ext})} \cdot \sqrt{\frac{1 + \varepsilon_{r_{eff}\_SW}}{2 \cdot \varepsilon_{r_{eff}\_SW}} - \frac{(h_1 + h_p)/h}{2 \cdot \varepsilon_{r_{eff}\_SW}}} \tag{4.50}$$

Considering the thicknesses $h_1 = 0.813$ mm, $h_p = 0.05$ mm and $h = 1.118$ mm and the lengths $L_{int} = 4.36$ mm and $L_{ext} = 5.7$ mm, the TE_{100} mode resonant frequency f_{0_SW} is equal to 11.32 GHz, which is comparable to the lowest frequency of 11.52 GHz obtained by EM simulation. A landing pad of thickness $h_p = 50\,\mu m$ and diameter 600 μm and a center-to-center distance of $s = 1$ mm was considered on the blind vias. An optimization of the SW-SIW antenna based on EM simulation was carried out in (Ho et al., 2021).

Photographs of the fabricated antennas are shown in Fig. 4.30. A reduction of 47% is obtained comparing the SW-SIW antenna with a conventional SIW fabricated on the same technology. Table 4.1 summarizes this comparison and highlights that the gain and effi-ciency of both SIW and SW-SIW antennas are comparable.

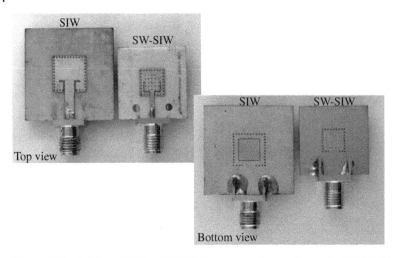

Figure 4.30 Fabricated SIW and SW-SIW antennas. Source: Ho et al., (2021)/with permission of IEEE.

Table 4.1 Comparison between SIW and SW-SIW.

	SIW	SW-SIW w/o inductive via	SW-SIW with inductive via
SIW cavity dimension (mm^2)	8.8 × 8.8	6.4 × 6.4	6.4 × 6.4
Measured – 10 dB bandwidth (MHz)	220	149	280
Measured gain (dBi)	5.5	5.7	4.8
Measured efficiency (%)	74	86	78
Polarization type	LP	LP	RHCP

4.5 SW-SIW in Metallic Nanowire Membrane Technology

Implementation of slow-wave integrated waveguides in advanced technologies such as silicon, glass, ceramic or even alumina membrane (MnM Technology) could be of high interest for the design of compact antenna arrays (including feeding network) or filters and diplexers at mm-waves. However, the integration of rectangular waveguides at mm-waves is still a subject of research and development with a limited number of realizations. The main reason for that is that integrating waveguides in conventional technologies back-end-of-line (BEOL) requires very large areas at least up to 300 GHz, resulting in costly solutions. Also, from the geometrical point-of-view, material thicknesses were chosen to provide a trade-off between cost and performance for integrated transmission lines such as microstrip, resulting in BEOL thickness from 5 to 10 μm for current advances Complementary Metal-Oxide-Semiconductor (CMOS)/Bipolar-CMOS (BiCMOS) technologies. On the other hand, waveguide thickness is the key parameter for achieving low-loss waveguides and 5 to 10 μm is clearly not sufficient to achieve high performance below at least 300 GHz.

The interposer technology addresses both thickness and surface limitations. It would, by construction, provide thicker dielectrics with larger areas, thus paving the way for integrated waveguides. On the other hand, if the occupied surface still needs to be reduced, slow-wave principles described in the previous sections could be once again applied.

The concept of PAF-SW-SIW was proposed by (Corsi, 2021) to improve SIW performance with a considerable increase in dimensions. This concept is based on the Air-filled-SIW (Parment et al., 2015) and the SW-SIW concept described in Section 4.4. In one hand, the Air-filled-SIW can present more that 3.5 times less losses than a classic SIW. On the other hand, its dimensions are much greater. Hence, to minimize the overall size while maintaining reduced losses, the slow-wave concept introduced in the previous section is explored through a partially-air-filled approach within the MnM technology presented in Chapter 3.

Figure 4.31 shows a 3D view the considered slow-wave SIW in MnM technology. The black dotted line in Fig. 4.31 outlines the boundaries of the PAF-SW-SIW structure. As in other slow-wave SIWs, the structure is divided in two parts. Here, the lower part is made of a nanoporous alumina membrane filled with copper nanowires (with heigh h_{nw}), acting as metallic posts. The top region of the waveguide is made of air with height h_{air}. These two regions are limited by metallic covers that are vertically connected by the copper nanowires in membrane. To create the air region in this slow-wave SIW, the top cover is suspended above the membrane and anchored at the side walls, as illustrated in Fig. 4.31.

A 50-μm-thick nanoporous anodic aluminum oxide (AAO) membrane from InRedox was used as substrate for the development of the PAF-SW-SIW structure. The membrane features nanopores with a diameter of 40 nm, spaced center-to-center at 107 nm intervals. These nanopores can be filled with copper to create nanowire-based Through Substrate Vias (TSVs) (Pinheiro et al., 2018). These TSVs are used to create the sidewalls of the waveguide, as illustrated in Fig. 4.31. The blind vias, used to create the slow-wave effect in (Niembro-Martin et al., 2014), are replaced by copper nanowires. In order to confine the electric field and create the slow-wave effect, an air cavity if formed between the top cover and the nanowire-filled substrate, which ensures the best dielectric properties. As illustrated in Fig. 4.38, to excite the first propagating mode and measure the performance of the PAF-SW-SIW, transitions from the grounded CPW to the PAF-SW-SIW were designed.

Figure 4.32 shows electromagnetic simulations of the electric and magnetic fields in the PAF-SW-SIW, using an equivalent anisotropic material, as explained above. It is possible to see in Fig. 4.32 that the electric field is restricted to the air region and the magnetic field occupies the entire waveguide, as explained in Chapter 1.

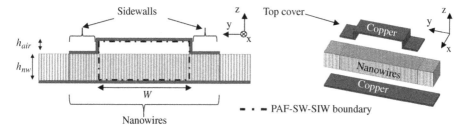

Figure 4.31 Cross-sectional view (left) and exploded 3-D view of the PAF-SW-SIW in MnM technology.

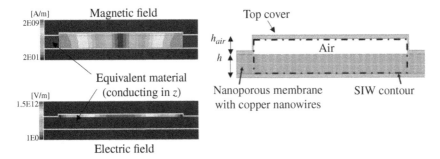

Figure 4.32 Magnetic and electric field distribution in the PAF-SW-SIW (left). Cross-section of the PAF-SW-SIW (right).

4.5.1 Effective Width and Cut-off Frequency

As presented in Section 4.3, the vertical conductivity of the metallic posts (nanowires in this technology) has a major impact in the performance of the waveguide. Fig. 4.33 shows the lateral confinement of the magnetic field for $\sigma_z = 10$ kS/m and $\sigma_z = 1$ MS/m.

It is possible to see the influence of the nanowire conductivity on the magnetic field penetration on the sidewall of the waveguide, where the penetration is greater for lower conductivities and dependent on the skin effect δ defined as (4.51).

$$\delta = \sqrt{\frac{2}{\mu_0 \cdot \omega \cdot \sigma_z}} \tag{4.51}$$

Since the electrical field is confined in the air region of the waveguide the conductivity of the nanowires does not influence the electric field.

This behavior of the magnetic field changes the effective width $W_{eff_{\sigma_z}}$ of the waveguide and leads to a variation of the cut-off frequency. Equation (4.52) accounts for the magnetic field penetration in the sidewalls of the waveguide and relates $W_{eff_{\sigma_z}}$ with the conductivity (expressed in terms of the skin effect), the membrane height h_{nw}, the air thickness h_{air}, and the width W of the waveguide.

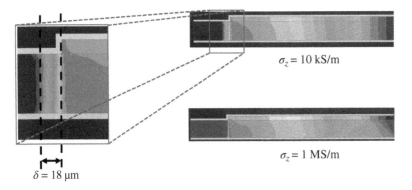

Figure 4.33 Influence of the nanowire conductivity on the lateral confinement of the magnetic field at 110 GHz in the PAF-SW-SIW with $W = 880$ μm, $h_{air} = 12$ μm and $h_{nw} = 50$ μm.

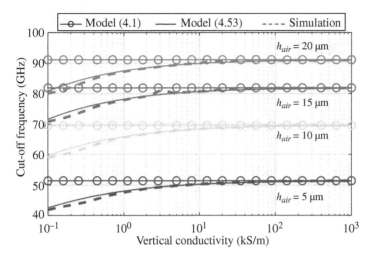

Figure 4.34 TE_{10} cut-off frequency of the PAF-SW-SIW versus the equivalent vertical membrane conductivity σ_z for air thicknesses from 5 to 20 μm ($W = 880$ μm, $h_{nw} = 50$ μm, $\varepsilon'_{r_2} = 9.7$, $\tan \delta = 0.015$).

$$W_{eff_{\sigma_z}} = W\sqrt{1 + \frac{2\delta}{W} \cdot \frac{h_{nw}}{h_{nw} + h_{air}}} \qquad (4.52)$$

For high conductivities, W equals $W_{eff_{\sigma_z}}$, since δ is very small ($\frac{2\delta}{W} \cdot \frac{h_{nw}}{h_{nw}+h_{air}}$ becomes negligible). In this case, the cut-off frequency of the SW-SIW is not affected. On the other hand, as the conductivity decreases, δ becomes relevant, consequently $W_{eff_{\sigma_z}}$ increases and the cut-off frequency decreases.

Figure 4.34 shows the cut-off frequency of the PAF-SW-SIW versus the equivalent vertical conductivity for air thicknesses ranging from 5 to 20 μm. The calculated f_c using equation (4.52), taking into account the nanowire conductivity, was compared to EM simulations considering the equivalent media described above. A good agreement between simulation and the presented model can be observed in Fig. 4.34. For comparison, the cut-off frequency using equation (4.1), without considering the conductivity of the nanowires, was also plotted.

A reduction of the cut-off frequency can be seen as the conductivity decreases, as expected. This behavior is more pronounced below 10 kS/m, above which, the cut-off frequency can be calculated with (4.1).

4.5.2 Losses due to Metallic Nanowires

The vertical conductivity of the nanowires in the PAF-SW-SIW also affects the attenuation constant α_{nw}. The nanowires in the sidewalls and the blind nanowires, illustrated in Fig. 4.35, have the same conductivity, but their contribution to the overall attenuation constant is distinct. They are named α_{sw} (sidewalls) and α_m (blind), respectively, in (4.53).

$$\alpha_{nw} = \alpha_{sw} + \alpha_m \qquad (4.53)$$

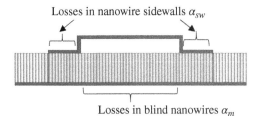

Losses in nanowire sidewalls α_{sw}

Losses in blind nanowires α_m

Figure 4.35 Illustration of the two sources of metallic losses due to metallic nanowires in the waveguide.

According to (Pozar, 2011), in a rectangular waveguide, the attenuation constant due to metallic loss in sidewalls is expressed by (4.54), which is the ratio between the dissipated power inside the walls P_{lsw} and the transmitted power P_{10} in the waveguide, represented by equations (4.55) and (4.56), where A_{10} is an arbitrary constant.

$$\alpha_{sw} = \frac{P_{lsw}}{2P_{10}} \tag{4.54}$$

$$P_{lsw} = R_s h_{nw} A_{10}^2 = \sqrt{\frac{\omega\mu_0}{2\sigma_z}} h_{nw} A_{10}^2 \tag{4.55}$$

$$P_{10} = \frac{\omega\mu_0 W_{eff}^3 \beta_{eff} A_{10}^2 h_{air} Re(\overline{SWF})}{4\pi^2} \tag{4.56}$$

The effective phase constant β_{eff} of the SW-SIW is given in equation (4.57) as a function of the effective width W_{eff} (4.52) and the complex relative effective dielectric constant $\overline{\varepsilon_{r_{eff}}}$ (4.36).

$$\beta_{eff} = Im\left(\sqrt{\left(\frac{\pi}{W_{eff}}\right)^2 - \overline{k}^2}\right) \tag{4.57}$$

with

$$\overline{k} = \frac{\omega}{c_0}\sqrt{\overline{\varepsilon_{r_{eff}}}}$$

Equation (4.58) expresses the attenuation constant α_m, which is related to the conductivity of the blind nanowires.

$$\alpha_m = Re\left(\sqrt{\left(\frac{\pi}{W_{eff}}\right)^2 - \overline{k}^2}\right) \tag{4.58}$$

Figure 4.36 presents EM simulations of the PAF-SW-SIW performed at 110 GHz as a function of the vertical conductivity. These simulations are compared with the calculated attenuation constants α_{sw} and α_m using (4.54) and (4.58), and show good agreement, especially for conductivities greater than 10 kS/m.

The extracted dielectric attenuation constant α_d from EM simulation is also presented in Fig. 4.36 and, it is nearly independent of the nanowire conductivity with very small values, when compared to α_{sw} and α_m. This was expected, because the electric field is confined in the air cavity.

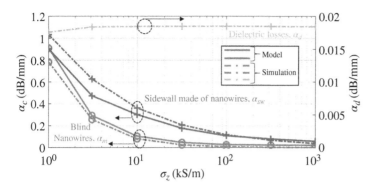

Figure 4.36 Attenuation constants (model and simulation) at 110 GHz of the PAF-SW-SIW versus the equivalent vertical membrane conductivity σ_z (with $h_{air} = 12\ \mu m$, $W = 880\ \mu m$, $\epsilon'_{r_2} = 9.7$, $\tan \delta = 0.015$ and $h_{nw} = 50\ \mu m$).

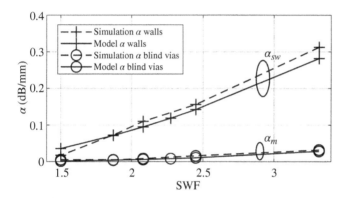

Figure 4.37 Attenuation constants α_{sw} and α_m versus the real part of the slow-wave factor $Re(\overline{SWF})$ for a vertical conductivity $\sigma_z = 10^2$ kS/m (with $f = 110$ GHz, $\epsilon'_{r_2} = 9.7$, $\tan \delta = 0.015$ and $h_{nw} = 50\ \mu m$).

Therefore, it is possible to conclude that the losses in the PAF-SW-SIW are mostly related to the metallic losses of the nanowire and the main contributor is the sidewall conductivity that has to be greater than to 10^2 kS/m to minimize α_{sw} and hence reduce the impact on the overall losses in the PAF-SW-SIW.

Figure 4.37 compares at 110 GHz and for $\sigma_z = 10^2$ kS/m, EM simulations and the calculated values for α_{sw} and α_m as a function of the real part of the complex slow-wave factor, $\overline{SWF}$ (from equation (4.27)). The SWF was changed by varying the height of the air cavity from 5 to 20 µm. As the SWF alters the cutoff frequency of the waveguide, these attenuation constants were calculated using different waveguide widths to ensure that their respective operating center frequencies f_0, classically defined as 1.5 times f_c, remain constant. These results present good agreement between EM simulations and the proposed model. Also, up to $SWF = 3.5$, it is possible to see that α_m has a negligible impact in the overall attenuation constant when compared to α_{sw}.

4.5.3 W-Band Implementation and Results

The theoretical analysis presented above emphasizes the significant influence of the membrane's equivalent vertical conductivity on the PAF-SW-SIW, particularly regarding losses. Consequently, meticulous care and consideration should be devoted to the growth of nanowires during the fabrication process to attain a low-loss structure.

Because on-wafer measurements at mm-waves involve the use of ground-signal-ground (GSG) probes, transition structures become essential. These structures are necessary to excite the waveguide and convert the TEM wave from the feeding lines into the TE_{10} wave propagating in the PAF-SW-SIW.

Figure 4.38 illustrates the realization of the PAF-SW-SIW, which uses 50-Ω grounded coplanar waveguides (GCPW), as proposed by (Songnan Yang et al., 2007), to improve the excitation of the TE_{10} in the PAF-WS-SIW. In the GCPW design, nanowires are utilized to connect the ground strips of the coplanar waveguide to the ground plane, as depicted in Fig. 4.38.

Two tapered transitions, as shown in Fig. 4.39, are used to establish the connection between the GCPW and the waveguide. The optimized transitions dimensions for a PAF-SW-SIW with a cut-off frequency of 75 GHz are detailed in Fig. 4.39. The first transition, denoted by its length L_t and width W_t, is designed to facilitate efficient electromagnetic wave propagation between the external 50-Ω GCPW and the air-suspended GCPW.

The second transition, characterized by its length L'_t and width W'_t, serves the purpose of transferring energy from the quasi-TEM GCPW to the TE_{10} propagation mode. An elliptical-shaped region, without nanowires, is incorporated at the initial part of the second taper to prevent a short circuit caused by nanowires between the central GCPW strip and the bottom ground plane. The dimensions of this transition were optimized through EM simulations, leading to the following design guidelines: (i) the width W'_t closely approximates one-third of the waveguide width W, (ii) the gap G'_t initially measures half of W'_t, and (iii) the length L'_t corresponds to one-quarter of the wavelength at the operational center frequency f_0 of the waveguide.

Technological constrains in the fabrication of the PAF-SW-SIW limited the air cavity height to 12 µm with a metal thickness of 5 µm. Considering an ideal equivalent vertical conductivity, the width of the nanowire sidewalls W_w was defined as 100 µm to account for

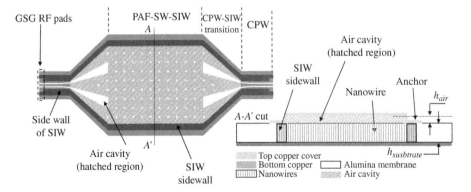

Figure 4.38 Top (left) and side-view (right) of the PAF-SW-SIW in the MnM technology

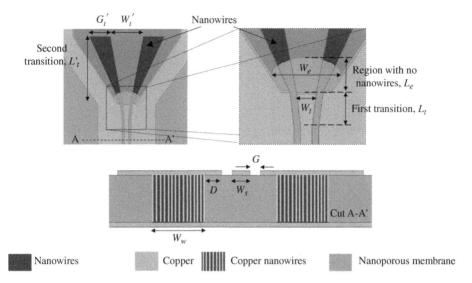

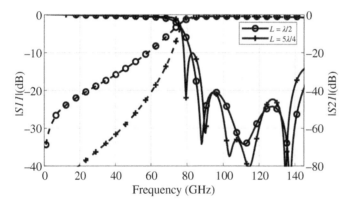

Figure 4.39 Transition between the GCPW feeding line and the PAF-SW-SIW (Corsi, 2021). Dimensions for a 75-GHz cut-off frequency: $L_t = 100, W_t = 60, L'_t = 531, W'_t = 272, G_t = 171, L_e = 97, W_e = 209. W_s = 35, G = 20, D = 27.5, W_w = 100$.

Figure 4.40 Simulated S-parameters of the PAF-SW-SIW with grounded coplanar waveguide (GCPW)-based transitions, for various lengths and a cut-off frequency of 75 GHz.

the skin effect (Fig. 4.33). A width of 880 µm was calculated for the PAF-SW-SIW for a cutoff frequency (TE_{10}) of 75 GHz.

Figure 4.40 presents the EM simulations of PAF-SW-SIW with the described transitions for different lengths ($L = \lambda/2$ and $L = 5\lambda/4$, where λ is the guided wavelength at the operating frequency f_0). These two lengths are required to extract the phase and attenuation constants of the waveguide. Based on the EM simulation, and for a return loss better than 20-dB, the relative bandwidth is equal to 39%. The insertion loss per transition is equal to 0.43 dB at $f_0 = 1.5 f_c$.

The fabrication process for the PAF-SW-SIW using MnM technology is outlined in Fig. 4.41. Initially, a 100-nm silicon dioxide mask layer (SiO_2) is deposited on the top

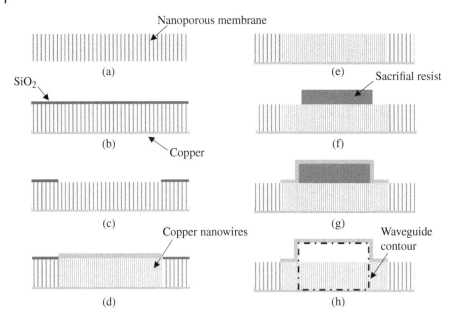

Figure 4.41 Fabrication steps of the PAF-SW-SIW on the MnM technology. a) deposition of SiO$_2$ mask layer on the front-side, b) coper seed layer deposition on the back-side, c) patterning of the SiO$_2$ film by photolithography and BEO etching, d) electrodeposition of copper nanowires, e) removal of excess copper by mechanical polishing, f) sacrificial resist deposition, g) copper deposition and thickening by electrodeposition and h) sacrificial resist removal.

surface of the anodic aluminium oxide (AAO) through reactive magnetron sputtering. Subsequently, a copper seed layer is sputter-deposited on the backside of the AAO. The SiO$_2$ is patterned using conventional photolithography and then etched with buffered oxide etch (BOE).

Nanowires are grown via electrodeposition through the nanopores of the AAO utilizing MacDermid Enthone's MacuSpec PPR 100 acid copper plating process. Electrodeposition continues until the nanowires reach the top side of the substrate. The top side of the membrane is then polished to align the nanowires flush with the membrane. A 12-μm-thick sacrificial layer of PMGI (Microchem) is deposited through spin coating and patterned.

A new copper layer is sputter-deposited on top of the PMGI to serve as a seed layer. Photoresist is employed as a mask in the final copper electrodeposition, leading to the growth of the top copper cover until it reaches a final thickness of 5 μm. Small holes measuring 20 × 20 μm are defined in the top cover to facilitate the etching of the sacrificial layer and the formation of the air cavity. These holes are considerably smaller than the wavelength and have no discernible impact on performance or electromagnetic field confinement. The copper on the backside is also thickened at this stage.

The remaining seed layer is etched away, and the PMGI sacrificial layer is removed. The fabricated PAF-SW-SIW structures are illustrated in Fig. 4.42.

Figure 4.42 displays a photograph of a processed one-inch membrane featuring various SIW structures, including the PAF-SW-SIW described above. A closer look at the waveguides is provided, showcasing examples with lengths of 5λ/4 and λ/2. At both ends of the waveguide,

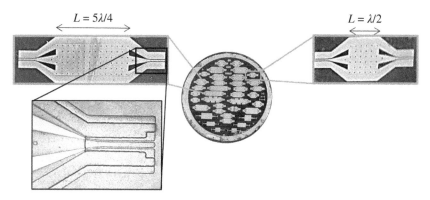

Figure 4.42 Fabricated PAF-SW-SIW on the MnM technology.

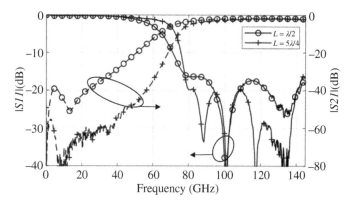

Figure 4.43 Measured S-parameters of PAF-SW-SIW of various lengths with GCPW transitions.

it is possible to observe the GCPW feeding lines and the tapered transitions. Additionally, the small holes used to aid in the release of the air cavity are visible in these views.

The measurement results for the 75-GHz waveguides up to 145 GHz were obtained using an ANRITSU MS4647B Vector Network Analyzer (VNA). These measurements were carried out with GSG probes and a Line-Reflect-Reflect-Match (LRRM) calibration was performed on a dedicated substrate.

The measured S-parameters of the PAF-SW-SIWs are presented in Fig. 4.43 for the two lengths considered in the simulation (as shown in Fig. 4.40). It is evident that, as anticipated, the cutoff frequency closely aligns with 75 GHz, with a maximum deviation of around 10 GHz observed between the different lengths. This indicates that the slow-wave effect achieved in these samples closely matches the intended design, meaning that the air layer's thickness is approximately the expected value. However, minor variations in the cutoff frequency (f_c) may be attributed to a potential sagging of the top metallic cover that is suspended in the air.

For the longest waveguide ($L = 5\lambda/4$), considering the GCPW transition, the measured insertion loss is 2.35 dB at 110 GHz. The return loss, represented by the S_{11} parameter, closely approaches the 20-dB level between 110 and 145 GHz.

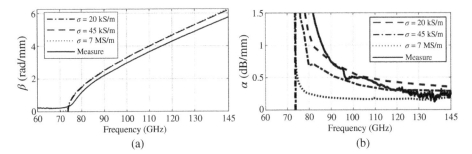

Figure 4.44 Propagation parameters of the PAF-SW-SIW (measurements versus EM simulations for various vertical conductivities σ_z): (a) phase constant β and (b) attenuation constant α.

Phase constant (β) and attenuation constant (α) extractions from the measured waveguides are conducted using a Transmission Line (TL) algorithm, which necessitates two structures of different lengths as described in (Souzangar & Shahabadi, 2009). The two PAF-SW-SIW structures presented in Fig. 4.42 are employed for this extraction, and the results are displayed in Fig. 4.44. Additionally, eigen-mode type simulations, as detailed in (Yousef et al., 2009), are performed using different values of vertical conductivity (σ_z) ranging from 20 kS/m to 7 MS/m.

The measured phase constant of the PAF-SW-SIW is found to be in good agreement with the expected results obtained in simulation.

The almost identical slope observed in the phase constant between measurement and simulation results indicates that the slow-wave factor, and consequently the air thickness, closely align with the designed values. As seen in the measured S-parameters in Fig. 4.43, slight variations in cutoff frequencies are noticeable between measurements and simulations, but these differences do not exceed 4%.

A reduction in losses is achieved with the PAF-SW-SIW in comparison to the classical SIW fabricated in the same technology (Bertrand et al., 2019a). Specifically, for the PAF-SW-SIW, the attenuation constant at the operating frequency ($f_0 = 1.5 \cdot f_c$) is approximately 0.4 dB/mm. This value is half the measured attenuation constant of the SIW counterpart, which is equal to 0.8 dB/mm.

However, the fabricated PAF-SW-SIWs exhibit higher losses than anticipated when compared to EM simulations assuming a vertical conductivity of 7 MS/m. To achieve better agreement with simulations, a lower vertical conductivity between 20 and 45 kS/m should be considered. This suggests that the quality of the nanowires used in fabrication may not be as good as expected.

4.5.4 SW-SIW Cavity Filters

Filters were designed using MnM technology to have a relative bandwidth of 10% around the center frequency of 112.5 GHz with a 5th-order Chebyshev response and residual ripple of 0.05 dB in the passband. A line filter without cross-coupling was considered in order to simplify this initial "proof-of-concept" design. The coupling topology of the SIW cavity resonators using inductive iris windows was chosen because it is commonly used in SIW for filter design due to its ease of fabrication and well known design methods (Chen &

Wu, 2008). However, the structure requires high-quality and precisely formed conductive walls so that the couplings and resonators are properly controlled. Thus, a second coupling structure, called a planar H-plane shifted cavity, was also used (Stephens et al., 2005). This is interesting because it does not require narrow conductive walls perpendicular to the wave propagation direction, which might be interesting considering the MnM technology. Moreover, it may be more resistant to fabrication variability than the iris topology presented above.

The design of these filters (iris and shifted cavity) is based on the method described in (Cameron et al., 2018). The filters are first defined according to the low-pass prototype and by calculating its normalized elements, which depend on the filter's order and response type (Matthaei et al., 1985).

The second step is to transform this first prototype into a bandpass filter, using series resonators and impedance inverters. As part of the design of a filter, the impedance of the inverters can be calculated as shown in (Matthaei et al., 1985), based on the normalized elements and the wavelength values at the low, high, and central cutoff frequencies of the filter. Details of the synthesis of both types of filters are presented in (Corsi, 2021).

The iris coupling filter is presented in the illustration of Fig. 4.45. It is composed of five resonators of equal lengths L_1, L_2, and L_3, separated from each other by coupling zones of width W_1, W_2, and W_3. As the structure is symmetrical, the resonators and couplings of the second half of the filter are considered to have the same dimensions as the first half. The design method consists first of modeling the coupling zones with an equivalent electrical circuit, then transforming this model into an impedance inverter, and finally plotting a characteristic impedance chart of the inverter as a function of the coupling width. This process allows the calculation of the lengths and width of the iris cavity filter.

After the synthesis of the filter, a 3D EM simulation (Ansys HFSS) is used to optimize the structure's response. A summary of the filter dimensions is available in Table 4.2 before and after an optimization. The small differences indicate that this design method is accurate and requires only a small amount of optimization work.

The H-plane shifted cavity filter is presented in Fig. 4.46, where the five resonators of lengths L_1, L_2, and L_3, and the dimension shifts S_1, S_2, and S_3, can be identified. The design method is different from that of iris filters in that it is entirely based on an analytical analysis of the cavity shift effect. This analysis is described in (Hunter, 1984) for a rectangular waveguide and has already been applied in the implementation of SIW-based filters at mm-waves (Aftanasar et al., 2002).

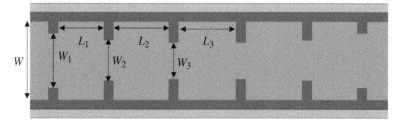

Figure 4.45 Top view of the iris coupling filter. Copper nanowires are present throughout the figure and the red areas represent the sidewalls of the waveguide.

Table 4.2 Summary of the dimensions of the iris coupling filter before and after an optimization phase.

	L_1 (μm)	L_2 (μm)	L_3 (μm)	W_1 (μm)	W_2 (μm)	W_3 (μm)
Before optimization	487.7	586.2	608.4	604.9	460.4	417.3
After optimization	481.7	587.4	608.2	606.2	451	408.6
Difference (%)	−1.2	0.2	0.03	0.21	2	0.02

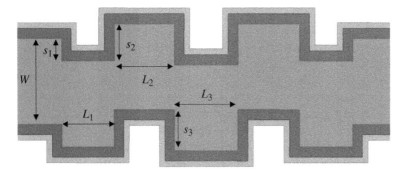

Figure 4.46 Top view of the 5th order H-plane cavity shifted filter. Copper nanowires are present throughout the figure and the red areas represent the sidewalls of the waveguide.

Table 4.3 Summary of the resonator dimensions and cavity shifted dimensions of the cavity shift filter.

	L_1 (μm)	L_2 (μm)	L_3 (μm)	S_1 (μm)	S_2 (μm)	S_3 (μm)
Before optimization	564.39	651.49	672.31	232.25	370.8	414.72
After optimization	555.76	643.42	669.81	230.37	380.32	419.84
Difference (%)	−1.53	−1.23	−0.37	−0.8	2.5	1.23

The resonator lengths are calculated using the same method as for the iris filter and the results are reported in Table 4.3 before and after EM optimization with Ansys HFSS, with a small deviation (maximum of 2.5%), which validates this design method for waveguide-based filters.

The simulated S-parameters of both iris and cavity shift filters are presented in Fig. 4.47 for a vertical conductivity of 6 MS/m and compared with the expected theoretical response without losses in black lines. The insertion loss for both filters is very close, and equal to 2.53 and 2.78 dB at the center frequency, respectively, for the iris and shifted cavity filters, and the rejection curves at both sides of the passband are identical. Next, the return loss for the cavity shift filter is less good as compared to the iris filter.

Next, the insertion loss of these two filters was studied as a function of the equivalent vertical conductivity of the membrane. The results are shown in Fig. 4.48 for conductivity

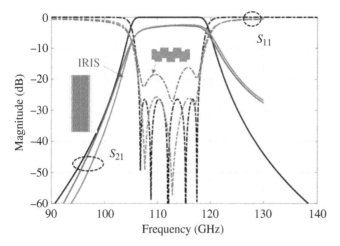

Figure 4.47 Optimized S_{21} (solid colored lines) and S_{11} (dotted lines) parameters of the iris and shifted cavity filters compared to the expected ideal response without losses (solid black lines). The vertical conductivity σ_z of the membrane is considered to be equal to 6 MS/m.

Figure 4.48 Insertion losses at 112.5 GHz for both filters as a function of the equivalent vertical conductivity of the membrane.

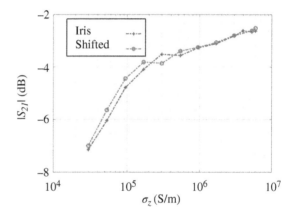

values ranging from 30 kS/m to 6 MS/m. The losses are equivalent for both structures regardless of the conductivity, but with a slight improvement for low conductivity values in the cavity shift filter. Note that the losses increase for vertical conductivity values below 100 kS/m, which is consistent with previous analysis (see Section 4.5), with a slope of 3.5 dB/dec as a function of σ_z. Beyond this value, the insertion loss varies less with conductivity, with a slope lower than 1 dB/dec.

References

Aftanasar, M. S., Young, P. R., & Robertson, I. D. (2002). Rectangular waveguide filters using photoimageable thick-film processing. *2002 32nd European Microwave Conference*, 1–4. https://doi.org/10.1109/EUMA.2002.339213

Bertrand, M., Corsi, J., Pistono, E., Kaddour, D., Puyal, V., & Ferrari, P. (2020). On the effect of field spatial separation on slow wave propagation. *IEEE Transactions on Microwave Theory and Techniques*, 68(12), 4978–4983. https://doi.org/10.1109/TMTT.2020.3023445

Bertrand, M., El Dirani, H., Pistono, E., Kaddour, D., Puyal, V., & Ferrari, P. (2019a). A 3-dB coupler in slow wave substrate integrated waveguide technology. *IEEE Microwave and Wireless Components Letters*, 29(4), 270–272. https://doi.org/10.1109/LMWC.2019.2900239

Bertrand, M., Liu, Z., Pistono, E., Kaddour, D., & Ferrari, P. (2015). A compact slow-wave substrate integrated waveguide cavity filter. *2015 IEEE MTT-S International Microwave Symposium*, 1–3. https://doi.org/10.1109/MWSYM.2015.7166800

Bertrand, M., Pistono, E., Kaddour, D., Puyal, V., & Ferrari, P. (2018). A filter synthesis procedure for slow wave substrate-integrated waveguide based on a distribution of blind via holes. *IEEE Transactions on Microwave Theory and Techniques*, 66(6), 3019–3027. https://doi .org/10.1109/TMTT.2018.2825403

Bertrand, M., Rehder, G. P., Serrano, A. L. C., Gomes, L. G., Pinheiro, J. M., Alvarenga, R. C. A., Kabbani, N., Kaddour, D., Puyal, V., Pistono, E., & Ferrari, P. (2019b). Integrated waveguides in nanoporous alumina membrane for millimeter-wave interposer. *IEEE Microwave and Wireless Components Letters*, 29(2), 83–85. https://doi.org/10.1109/LMWC.2018.2887193

Bozzi, M., Pasian, M., Perregrini, L., & Wu, K. (2007). On the losses in substrate integrated waveguides. *2007 European Microwave Conference*, 384–387. https://doi.org/10.1109/EUMC .2007.4405207

Cameron, R. J., Kudsia, C. M., & Mansour, R. R. (2018). *Microwave Filters for Communication Systems*. John Wiley & Sons, Inc., 457–483. https://doi.org/10.1002/9781119292371

Cassivi, Y., Perregrini, L., Arcioni, P., Bressan, M., Wu, K., & Conciauro, G. (2002). Dispersion characteristics of substrate integrated rectangular waveguide. *IEEE Microwave and Wireless Components Letters*, 12(9), 333–335. https://doi.org/10.1109/LMWC.2002.803188

Chen, C.-J., & Chu, T.-H. (2010). Design of a 60-GHz substrate integrated waveguide butler matrix—a systematic approach. *IEEE Transactions on Microwave Theory and Techniques*, 58(7), 1724–1733. https://doi.org/10.1109/TMTT.2010.2050097

Chen X-P, & Wu, K. (2008). Accurate and efficient design approach of substrate integrated waveguide filter using numerical TRL calibration technique. *2008 IEEE MTT-S International Microwave Symposium Digest*, 1231–1234. https://doi.org/10.1109/MWSYM.2008.4633281

Collin, R. E. (1990). *Field Theory of Guided Wave* (Vol. 5). John Wiley & Sons, 411–471.

Corsi, J. (2021). *Guides à ondes lentes intégrés en technologie interposeur: applications aux filtres passe-bande aux longueurs d'onde millimétriques*. Université Grenoble Alpes, 73–81.

Corsi, J., Rehder, G. P., Gomes, L. G., Bertrand, M., Serrano, A. L. C., Pistono, E., & Ferrari, P. (2020). Partially-air-filled slow-wave substrate integrated waveguide in metallic nanowire membrane technology. *IEEE MTT-S International Microwave Symposium Digest*, *2020-August*. https://doi.org/10.1109/IMS30576.2020.9223955

Deslandes, D. (2005). *Etude et developpement du guide d'ondes integre au substrat pour la conception de systemes en ondes millimetriques*. Université de Montréal.

Deslandes, D., & Wu, K. (2003). Single-substrate integration technique of planar circuits and waveguide filters. *IEEE Transactions on Microwave Theory and Techniques*, 51(2), 593–596. https://doi.org/10.1109/TMTT.2002.807820

Franc, A.-L., Podevin, F., Cagnon, L., Ferrari, P., Serrano, A., & Rehder, G. (2012). Metallic nanowire filled membrane for slow wave microstrip transmission lines. *2012 International Semiconductor Conference Dresden-Grenoble (ISCDG)*, 191–194. https://doi.org/10.1109/ISCDG.2012.6360022

Franck, P., Baillargeat, D., & Tay, B. K. (2012). Mesoscopic model for the electromagnetic properties of arrays of nanotubes and nanowires: a bulk equivalent approach. *IEEE Transactions on Nanotechnology*, 11(5), 964–974. https://doi.org/10.1109/TNANO.2012.2209457

Ho, A. T., Pistono, E., Corrao, N., & Ferrari, P. (2021). Circular polarized square slot antenna based on slow-wave substrate integrated waveguide. *IEEE Transactions on Antennas and Propagation*, 69(3), 1273–1282. https://doi.org/10.1109/TAP.2020.3030933

Hunter, J. D. (1984). The displaced rectangular waveguide junction and its use as an adjustable reference reflection. *IEEE Transactions on Microwave Theory and Techniques*, 32(4), 387–394. https://doi.org/10.1109/TMTT.1984.1132687

Janaswamy, R., & Schaubert, D. H. (1986). Characteristic impedance of a wide slotline on low-permittivity substrates (short paper). *IEEE Transactions on Microwave Theory and Techniques*, 34(8), 900–902. https://doi.org/10.1109/TMTT.1986.1133465

Jin, H., Wang, K., Guo, J., Ding, S., & Wu, K. (2016). Slow-wave effect of substrate integrated waveguide patterned with microstrip polyline. *IEEE Transactions on Microwave Theory and Techniques*, 64(6), 1717–1726. https://doi.org/10.1109/TMTT.2016.2559479

Jin, H., Zhou, Y., Huang, Y. M., & Wu, K. (2017). Slow-wave propagation properties of substrate integrated waveguide based on anisotropic artificial material. *IEEE Transactions on Antennas and Propagation*, 65(9), 4676–4683. https://doi.org/10.1109/TAP.2017.2726688

Kim, D.-Y., Lee, J. W., Lee, T. K., & Cho, C. S. (2011). Design of SIW cavity-backed circular-polarized antennas using two different feeding transitions. *IEEE Transactions on Antennas and Propagation*, 59(4), 1398–1403. https://doi.org/10.1109/TAP.2011.2109675

Mangan, A. M., Voinigescu, S. P., Yang, M.-T., & Tazlauanu, M. (2006). De-embedding transmission line measurements for accurate modeling of IC designs. *IEEE Transactions on Electron Devices*, 53(2), 235–241. https://doi.org/10.1109/TED.2005.861726

Matthaei, G. L., Young, L., & Jones, E. M. T. (1985). *Microwave Filters, Impedance-Matching Networks, and Coupling Structures*. (Artech House), 83–162.

Niembro-Martin, A., Nasserddine, V., Pistono, E., Issa, H., Franc, A.-L., Vuong, T.-P., & Ferrari, P. (2014). Slow-wave substrate integrated waveguide. *IEEE Transactions on Microwave Theory and Techniques*, 62(8), 1625–1633. https://doi.org/10.1109/TMTT.2014.2328974

Parment, F., Ghiotto, A., Vuong, T.-P., Duchamp, J.-M., & Wu, K. (2015). Air-filled SIW transmission line and phase shifter for high-performance and low-cost U-Band integrated circuits and systems. *Global Symposium on Millimeter-Waves (GSMM)*, 1–3. https://doi.org/10.1109/GSMM.2015.7175444

Pinheiro, J. M., Rehder, G. P., Gomes, L. G., Alvarenga, R. C. A., Pelegrini, M. V., Podevin, F., Ferrari, P., & Serrano, A. L. C. (2018). 110-GHz through-substrate-via transition based on copper nanowires in alumina membrane. *IEEE Transactions on Microwave Theory and Techniques*, 66(2), 784–790. https://doi.org/10.1109/TMTT.2017.2763142

Pozar, D. M. (2011). *Microwave Engineering*. John Wiley & sons, 110–121.

Salehi, M., & Mehrshahi, E. (2011). A closed-form formula for dispersion characteristics of fundamental SIW mode. *IEEE Microwave and Wireless Components Letters*, 21(1), 4–6. https://doi.org/10.1109/LMWC.2010.2088114

Serrano, A. L. C., Franc, A.-L., Assis, D. P., Podevin, F., Rehder, G. P., Corrao, N., & Ferrari, P. (2014). Modeling and characterization of slow-wave microstrip lines on metallic-nanowire-filled-membrane substrate. *IEEE Transactions on Microwave Theory and Techniques*, 62(12), 3249–3254. https://doi.org/10.1109/TMTT.2014.2366108

Souzangar, P., & Shahabadi, M. (2009). Numerical multimode thru-line (TL) calibration technique for substrate integrated waveguide circuits. *Journal of Electromagnetic Waves and Applications*, 23(13), 1785–1793. https://doi.org/10.1163/156939309789566969

Stephens, D., Young, P. R., & Robertson, I. D. (2005). Millimeter-wave substrate integrated waveguides and filters in photoimageable thick-film technology. *IEEE Transactions on Microwave Theory and Techniques*, 53(12), 3832–3838. https://doi.org/10.1109/TMTT.2005.859862

Uchimura, H., Takenoshita, T., & Fujii, M. (1998). Development of a "laminated waveguide". *IEEE Transactions on Microwave Theory and Techniques*, 46(12), 2438–2443. https://doi.org/10.1109/22.739232

Xu, F., & Wu, K. (2005). Guided-wave and leakage characteristics of substrate integrated waveguide. *IEEE Transactions on Microwave Theory and Techniques*, 53(1), 66–73. https://doi.org/10.1109/TMTT.2004.839303

Songnan Yang, Elsherbini, A., Song Lin, Fathy, A. E., Kamel, A., & Elhennawy, H. (2007). A highly efficient Vivaldi antenna array design on thick substrate and fed by SIW structure with integrated GCPW feed. *2007 IEEE Antennas and Propagation Society International Symposium*, 1985–1988. https://doi.org/10.1109/APS.2007.4395912

Yousef, H., Cheng, S., & Kratz, H. (2009). Substrate integrated waveguides (SIWs) in a flexible printed circuit board for millimeter-wave applications. *Journal of Microelectromechanical Systems*, 18(1), 154–162. https://doi.org/10.1109/JMEMS.2008.2009799

Index

Slow-Wave Microwave and mm-Wave Passive Circuits, First Edition. Edited by Philippe Ferrari, Anne-Laure Franc, Marc Margalef-Rovira, Gustavo P. Rehder, and Ariana Lacorte Caniato Serrano.
© 2025 John Wiley & Sons Ltd. Published 2025 by John Wiley & Sons Ltd.